Nuclear Cultures

Nuclear Cultures: Irradiated Subjects, Aesthetics and Planetary Precarity aims to develop the field of nuclear humanities and the powerful ability of literary and cultural representations of science and catastrophe to shape the meaning of historic events. Examining multiple discourses and textual materials, including fiction, poetry, biographies, comics, paintings, documentary and photography, this volume will illuminate the cultural, ecological and social impact of nuclearization narratives. Furthermore, this text explores themes such as the cultures of atomic scientists, the making of the bomb, nuclear bombings and disasters, nuclear aesthetics and art, and the global mobilization against nuclearization. *Nuclear Cultures* breaks new ground in the debates on "the nuclear" to foster the development of nuclear humanities, its vocabulary and methodology.

Pramod K. Nayar, recipient of the Visitor's Award for Best Research in Arts, Humanities and Social Sciences from the President of India, is Professor in the Department of English at the University of Hyderabad, India. He is the author of books in literary-cultural studies, with primary interests in English colonial writings, posthumanism and, in the past decade, human rights and literature. His newest books include *Alzheimer's Disease Memoirs* (2021), *The Human Rights Graphic Novel* (2020), *Ecoprecarity* (2019), *Indian Travel Writing in the Age of Empire* (2020) and *Essays in Celebrity Culture* (2021). Nayar holds the UNESCO Chair in Vulnerability Studies at the University of Hyderabad.

Routledge Studies in World Literatures and the Environment
Series Editors: Scott Slovic and Swarnalatha Rangarajan

For more information about this series, please visit: www.routledge.com/
Routledge-Studies-in-World-Literatures-and-the-Environment/book-series/
ASHER4038

Nuclear Cultures
Irradiated Subjects, Aesthetics and Planetary Precarity

Pramod K. Nayar

Routledge
Taylor & Francis Group

NEW YORK AND LONDON

First published 2023
by Routledge
605 Third Avenue, New York, NY 10158

and by Routledge
4 Park Square, Milton Park, Abingdon, Oxon, OX14 4RN

Routledge is an imprint of the Taylor & Francis Group, an informa business

© 2023 Pramod K. Nayar

ISBN: 978-1-032-18384-8 (hbk)
ISBN: 978-1-032-18391-6 (pbk)
ISBN: 978-1-003-25429-4 (ebk)

DOI: 10.4324/9781003254294

Typeset in Sabon
by Newgen Publishing UK

Contents

Figures

Preface

This book deals with 'nuclear cultures' and cuts across genres, media and themes. As a consequence, it is by necessity selective in the texts and themes chosen from the vast archive of nuclear texts. It is more than likely that texts that reinforce or contest the arguments made here figure in the archives, but that is a risk immanent to a project such as this.

There are extensive quotations from the primary materials – employed to prove rather than to illustrate – and the variety of these materials is symbolic of the quantum and range in nuclear cultures.

The work does not account for all the debates around nuclear power and nuclearization (e.g., that of India), opting instead for specific themes and features, such as the nuclear subject and its formation or aesthetics.

Parts of this book were written when the Doomsday Clock shows 100 seconds to midnight, capturing the threat to the world from Russia's Ukrainian misadventure, and thus seems apposite, and unfortunate.

June 2022
Hyderabad, India

Acknowledgements

To Michelle Salyga, Senior Editor, Literature, at Routledge New York, for receiving with great enthusiasm the first email carrying the idea for a book on the 'nuclear humanities', coordinating the proposal review with great speed, and all subsequent editorial encouragement that has enabled this book, my many thanks.

To Bryony Reece, editorial assistant at Routledge, for her immediate responses to queries, this book owes much.

My productive life is led in a small – some would say, tiny – inclusion zone, and it is a privilege to describe and thank its residents.

My parents and parents-in-law, and Nandini and Pranav, continue their rock-solid support for all that I do, with the latter often bearing the brunt of my odd sleep patterns and working modes. Pranav's interest in the *Chernobyl* series ensured I did not miss its finer points, which he inventoried in detail. But for this supportive ecosystem – a nuclear family! – work would simply not get done.

There are those who unfailingly track my working life, state of mind and health with cheer, affection and interest, to whom, this book, like the others before, owes a huge debt: Neelu, Bhalla, Savitha, Archana, Shubha, Ajeet, Ibrahim, Naveed, Vaishali, Soma, Haneef, Kishore, Manoj, Premlata and Kulsoom. Many of them are also responsible for the memes, jokes and funnies that populate my WApp daily – those not responsible for sending them to me are of course the helpless recipients of such explosive material (from me). Neelu, with her recent habit of sending snippets of songs, continues to engage me in debates about the respective merits of Rafi and Kishore – as a welcome distraction.

From a distance, Molly (Chechu) watches over me, and to this generous soul, I owe much. And then some.

The smoothness with which Nandana, now a seasoned campaigner when it comes to my work, gets me thinking, particularly along lines I do *not* want to think about, but which in the long run always aid my work, is an indispensable component of my writing life. For her interjections and 'mmmm . . . I am not sure what you mean by this' messages, and her affection and love of over 25 years: how do I even begin to thank her?

Professor Narayana Chandran, completely inadvertently on occasion, deflects my thinking onto some cleverly concealed paths – which when explored, circumspectly at first, but with more confidence later, open up vistas and prospect views. KNC remains inspirational with not just his monumental scholarship but also for his generosity.

Shubha Nagarkar produced, magically it seemed to me, essays from databases within minutes. To her affectionate friendship, energy and focus, I am extremely grateful.

In more recent times, conversations with Sheela Suryanarayanan have energized the work atmosphere, and I gratefully acknowledge this. It is also a delight to be able to thank Saradindu Bhattacharya, who brings several of his ideas for conversations.

To Kailash, Sharmistha, Debashis, warriors who co-survived a certain kind of fallout, thank you.

To Zahid for the afternoon chats-and-chai; Ashish whose old-world hospitality is charming (like his fascination for Nilufer biscuits, introduced with '*ye ek try karo*'); Krishna Ram and Jawahar *saab* – thank you for the many impromptu meetings.

I thank Bhim Singh of the Department of Hindi for providing me, at remarkably short notice, a useable edition of Agyeya's 'Hiroshima'.

And Anna. My First Reader for book manuscripts and essays, messaged: 'I liked reading it. A lot'. When the script returned to me, it had its usual notations, underlined passages and corrections. While she toils valiantly to straighten my prose, Anna also shapes the words in my head. That some of them do not translate into stylish sentences and prepositional perfection is entirely my fault. To Anna's love and the understanding that surpasses belief, I owe this book too.

For the canine friends who, literally, dog my heels in the mornings near my residence, and extract pats and belly-rubs as toll, I owe the pleasure of the walk (undertaken by hopping clumsily from side to side to avoid their feet). To the ones who shower their affections at the Humanities building, Anna and I owe the joyous start of our working day. To Ramesh, the Department of English Monarch, now an Insta-star and mascot on our Indian Writing in English Online OER (https://indianwritinginenglish. uohyd.ac.in), the solid presence in our classes, offices, meetings and events, I ought to compose some dog/gerel, but shall just satisfy myself with an expression of affectionate gratitude. Take a bow, Ramesh!

Parts of this book appeared in shorter, even different versions in journalism ('When Art Goes Nuclear', *Telangana Today*, 21 January 2022, 'One World or None: Nuclear Landscapes', *Telangana Today*, 3 April 2022, 'Our Failure to Stop Our Destruction', *Telangana Today*, 5 April 2022) and talks. Sections of the chapter, 'The A-List' were delivered as the keynote address titled 'The A-List: Atomic Scientists and the Nuclear Humanities' at the conference 'Cultural Trajectories through Language, Literature, and Media' of the Department of Humanities and Social

Sciences at the LNM Institute of Information Technology, Jaipur (29 April 2022). I am grateful to Usha Kanoongo for inviting me to speak at this conference. Chapter 5 will appear in segmented form as ' "One World or None": Planetary Nuclear Precarity and Anti-nuclear Cosmopolitanism' in *Planetary Precarity*, edited by Janet Wilson, Om Dwivedi and Barbara Schmidt-Haberkamp. The chapter on nuclear scientists will appear, also in modified and truncated form, as in *Life Writing as World Literature*, edited by Ioana Luca and Helga Lenart-Cheng for Bloomsbury.

Sections of the chapter on planetary precarity and anti-nuclear cosmopolitanism were enunciated as a plenary talk, 'Nuclear Colonialism: Indigeneity, Landscape and the Postcolonial' delivered at the conference, 'The Mirror of Life: Representations of History, Culture and Nature in Literature', St. Xavier's College, Palayamkottai, Tamil Nadu (7 May 2022), thanks to the invitation from D. Jockim. The arguments about *Trinity: A Graphic History of the First Atomic Bomb* were first rehearsed in an essay, 'Graphic Science: *Trinity* and the Art of the Atomic Bomb' published in *Rhizomes* 34 (2018).

I gratefully acknowledge Peter Goin and Ian McLaren-the Imperial War Museum for permissions to reproduce their photograph and anti-nuke poster, respectively.

I am grateful to the Professional Development Fund of the Institution of Eminence project of the University of Hyderabad.

1 Introduction

Nuclear Cultures

Masui Ibuse in *Black Rain* (1969) speaks of the efforts of a Hiroshima survivor, Shigematsu, at writing a journal of the bombing. The conversation between him and his wife, Shigeko, goes like this:

> "About your journal of the bombing—" she began. "You're going to present it to the library for posterity, aren't you? That's right, isn't it?"
> "That's right. The headmaster asked me to. It's my piece of history."
> "Then you ought to take more care over it. Why don't you write it out with writing-brush ink instead of ordinary pen ink? Writing in pen ink gradually fades away with time, doesn't it?"
> "Don't be silly. It may fade a little, but not as much as all that."
> (40)

After this conversation, in which his wife draws attention to the need to ensure durability of the memorializing in terms of the quality of ink and paper, Shigematsu reaches a decision:

> Shigematsu decided to rewrite his "Journal of the Bombing" using a brush and Chinese ink. He would have Shigeko copy out again the part already written with a pen, and would himself go on with the rest on Japanese-style writing paper, using a brush.
> He had been so thirsty that day. He would have given anything for a drink of water. He had turned a tap by the roadside, and steaming hot water had come out, too hot to drink directly, too hot to cup in the hands . . . His head full of such memories, he took up his brush, and set to work.
> (44)

Ibuse's is a comment on memorializing, on the physical act of writing an eyewitness and experiential account of the atomic bombing of the Japanese, on the materiality of memory-making and the mode of writing a 'personal piece of history'. It is an instance of the nuclear cultures this

DOI: 10.4324/9781003254294-1

book is interested in, and nuclear cultures are a subset of what has been termed the 'nuclear humanities'.

In their pioneering volume, *Reimagining Hiroshima and Nagasaki: Nuclear Humanities in the Post-Cold War* (2018), N.A.J. Taylor and Robert Jacobs speak of the urgent need to gather 'a variety of perspectives to gain moral and political insights on the full range of vulnerabilities—such as emotional, bodily, cognitive, and ecological—that pertains to nuclear harm' (3). These perspectives along with the mnemonics around Hiroshima and Nagasaki – the 'recollection, memorialization, and commemoration of Hiroshima and Nagasaki by officials and states, but also ordinary people's resentment, suffering, or forgiveness' (2) – constitute the broad transdisciplinary field of the Nuclear Humanities. The aim of their volume, they argue, is to examine the 'mediation between artist and critics who are grappling with what nuclear culture and arts can do' (9). Artists, critics and authors contribute to our thinking about nuclear *harm* in particular, and thereby mediate our sense of the nuclear itself. They also construct the human and the non-human as nuclear subjects, located at the epicentre of any nuclear event.

The Nuclear Humanities in this book is concerned with specific forms of nuclear cultures in which the human, non-human but also the non-living exist under the constant condition of nuclear threat, harm and possible extinction, and commentators have noted this state of constant threat (the Doomsday Clock is one symbol of this state). There is, argues Jean-Luc Nancy about Fukushima, the 'close and brutal connections between a seismic quake, a dense population, and a nuclear installation' (2015: 30). Nuclear cultures capture most of these aspects: Nature, science and technology, the human and the non-human, all mediated, irradiated by nuclear energy.

Nuclear Cultures

This book is interested in four aspects of 'nuclear cultures': the construction of the persona of the atomic scientist, the nature of the nuclear (human) subject, the aesthetics of catastrophe and the anti-nuclear discourses that foreground the theme of planetary precarity.

Nuclear cultures of planetary precarity manifest in literary texts, popular fiction and film, comics and biographical narratives of scientists. There are eyewitness–survivor accounts, drawings, collections of photographs, poetry, oral testimonies and interviews, plays and graphic novels about nuclear precarity, produced by victims from Hiroshima–Nagasaki, but also 'downwinders' from Nevada, the Aboriginal residents of Australia and the Pacific Islands. There are accounts and fiction set in and around the Chernobyl and Fukushima disasters. Documentary films and anti-nuclear artwork also make up a significant part of the oeuvre of nuclear cultures, although I do not examine specific protests and protest sites such as Koodankulam (Tamil Nadu, India) or Finland.

In what follows, I offer a brief survey of the texts around nuclear cultures, and many of these texts will figure in the later chapters.

Literary Fiction

From the early proleptic text, H.G. Wells' *The World Set Free* (1914) through Nevil Shute's *On the Beach* (1957), a nuclear apocalypse has provided fodder for numerous literary and popular fiction texts. Japanese fiction on the bombings has appeared steadily for many years (Treat 1996). Arguably, this strand – nuclear disaster/apocalypse – in nuclear cultures has been the strongest. Wells' novel anticipated the role uranium would play in the century to come. In fact, in its early sections, a professor of Chemistry waxes eloquent about these elements:

> [with the] use [of] this uranium and thorium; not only should we have a source of power so potent that a man might carry in his hand the energy to light a city for a year, fight a fleet of battleships, or drive one of our giant liners across the Atlantic; but we should also have a clue that would enable us at last to quicken the process of disintegration in all the other elements, where decay is still so slow as to escape our finest measurements. Every scrap of solid matter in the world would become an available reservoir of concentrated force . . . The energy we need for our very existence, and with which Nature supplies us still so grudgingly, is in reality locked up in inconceivable quantities all about us . . .
>
> (unpaginated)

The quest for unlimited and massive power, Wells predicts, would lead humanity, perhaps precariously, into Nature itself, and the atom. Even before Trinity, in the 1900–1935 period, numerous novelists anticipated the role of uranium, radiation and other means of atomic power – so many, in fact, that the MIT Press set up a series of reprints of such work under the series, 'The Radium Age'.[1]

Post-Trinity and Hiroshima–Nagasaki, such fiction grew in numbers. Philip Wylie's *Tomorrow!* (1954) evoked a nuclear war scenario, and the attendant chaos in small-town America. Shute's novel envisioned the end of the world in a nuclear apocalypse. Pat Frank's *Alas, Babylon* (1959) focused on the post-nuclear survival of the few (a classic apocalyptic scenario). When the novel ends, the discussion revolves around who actually won the war:

> Randy said, "Paul, there's one thing more. Who won the war?"
> Paul put his fists on his hips and his eyes narrowed. "You're kidding! You mean you really don't know?"
> "No. I don't know. Nobody knows. Nobody's told us." "We won it. We really clobbered 'em!" Hart's eyes lowered and his arms drooped. He said, "Not that it matters."

The engine started and Randy turned away to face the thousand-year night.

(unpaginated)

Set in the Cold War period, Kurt Vonnegut Jr.'s black satire, *Cat's Cradle* (1963) served as an allegory for the nuclear threat. Peter George's *Dr Strangelove* (1964) went on to become a major film. Towards the latter decades of the twentieth century and the early twenty-first, literary interest in nuclear power plants, nuclear apocalypse and disasters has continued. David Mitchell's *Cloud Atlas* (2004) references nuclear power plant safety. In the post-Fukushima world, hundreds of Japanese stories have appeared dealing with the 'triple-disaster' (DiNitto 2019, Arribert-Narce 2021). The Chernobyl survivors and their life in the so-called Exclusion Zone, to which Marusia and other old women return, is the subject of Irene Zabytko's *The Sky Unwashed* (2012). Zabytko captures the culture of secrecy that was integral to the events of Chernobyl (and its effects): the official proscription of knowledge, the silence over the history of earlier smaller accidents, negligence over reactor design, etc. ('Zosia knew about the common small fires and short circuits and power blackouts that regularly occurred at the Chornobyl plant', writes Zabytko at one point, unpaginated). More recent works, such as James George's epic nuclear novel, *Ocean Roads* (2007), brings together America, Japan and New Zealand together, demonstrating an intercontinental and intergenerational experience of nuclear harm.

Survivor/Eyewitness Accounts

A vast amount of survivor and eyewitness accounts, in the form of memoirs, compilations and collections, principally from the Hiroshima–Nagasaki bombings, exists. Kenzaburo's *Hiroshima Notes*, based on a later visit to Hiroshima, appeared in English translation in 1981. Michihiko Hachiya's *Hiroshima Diary*, an account of his days as a survivor and a physician, appeared in 1983. Richard Minear compiled eyewitness accounts in *Hiroshima: Three Witnesses* (1991), as did Akiko Mizue in *Hiroshima: Survivors' Testimonies* (2020). *The Bells of Nagasaki*, the memoir of the celebrity survivor-physician, Takashi Nagai (1984), Arata Osada's *Children of the A-Bomb: Testament of the Boys and Girls of Hiroshima* (2015) and Kyoko and Selden's *The Atomic Bomb: Voices from Hiroshima and Nagasaki* (2015) constitute invaluable compilations of eyewitness writings and interviews. Nagai's text concluded with the controversial statements about Nagasaki having been 'chosen' for the sacrifice, and immediately attracted global attention.

From other regions harmed to lesser or greater degrees by nuclear tests and experiments but have received comparatively less visibility, we now have historical and survivor narratives: the Marshall Islands, the Nevada-Utah 'downwinders', the Maralinga tests, etc. (Dibblin 1990,

Lennon 2011, Burgess 2012, Tynan 2016, respectively). Lennon's memoir was the first to document the effect of nuclear testing on the lives of the Aboriginal dwellers of the Maralinga region. Dibblin's interviews with the Marshall Island survivors – displaced from the region to enable American nuclear tests – demonstrates the long-term effects of the radiation that pervaded the region. The 1954 *Lucky Dragon* incident has its own survivor account in the form of Ōishi Matashichi's *The Day the Sun Rose in the West: Bikini, the Lucky Dragon, and I* (2011).

Reportage and Histories

Western journalist reportage on the bombings appeared very early (Burchett, Hersey). Hersey's best-selling *Hiroshima* examines the events through the accounts of six survivors. After graphic descriptions of the devastation, Hersey concludes with the shift of status from 'survivors' to *hibakusha*, that the Japanese themselves opted for. George Weller's *First into Nagasaki* was originally drafted as despatches, four weeks after the bombing, and were heavily censored by the Occupation forces stationed in Japan. As Walter Cronkite's Foreword tells us:

> Weller salvaged his carbon copy but, in his subsequent travels to many corners of our troubled globe, the copy disappeared. His son, an honored writer in his own right, has only recently uncovered it.
>
> (unpaginated)

'Insider' and/or survivor/eyewitness accounts of some of the worst nuclear disasters like Chernobyl, the 'clean-up' operations in Chernobyl and Fukushima, have appeared since the 1990s (Tatsuta 2017), alongside journalistic accounts based on primary materials and interviews. Grigoriy Medvedev in *Chernobyl Notebook* (1987) traces a history of negligence and politics around Chernobyl. Svetlana Alexievich in *Chernobyl Prayer* (2016) interviews numerous officials, survivors and the afflicted, demonstrating the lingering effects of radiation in the region. Histories, distilled from hundreds of interviews with officers, soldiers, engineers and others, but also through access to hitherto unavailable documents, have appeared recently: Adam Higginbotham's *Midnight in Chernobyl* (2017), Serhii Plokhy's *Chernobyl* (2018) and Kate Brown's *Manual For Survival* (2019). Post-disaster accounts of these 'Exclusion Zones' have also appeared in print (Richter 2020), with some focusing on specific topics such as the return of wildlife to the region (Johnson 2015).

Historians have traced the politics and process of the making of the atomic bomb, many of which include eyewitness reports of events like Trinity, and the nuclear arms race (Groves 1962, Boyer 1998, Kelly 2007, Rhodes 2012, Cordle 2017, Laurence 2017). Works like Rhodes' Pulitzer winning *The Making of the Atomic Bomb* offers a wealth of detail from the archives, lives and events, thereby helping us trace a public chronicle of nuclearity.

Visual Texts

Photographs from Hiroshima–Nagasaki in official collections such as the Manhattan Engineer District's three-part *Photographs of the Atomic Bombings of Hiroshima and Nagasaki* (1945) have captured aerial views of the destruction. Panoramic shots show demolished buildings, bridges and acres of ruins. These panoramic shots with their absence of the *human* ruins are offset by Japanese photographs in collections such as Donald Goldstein et al.'s *Rain of Ruin: A Photographic History of Hiroshima and Nagasaki* (1999) and *Flash of Light, Wall of Fire: Japanese Photographs Documenting the Atomic Bombings of Hiroshima and Nagasaki* (2020). These focus on the injured, broken, burned and dying Japanese. Then there are the photographs of the abandoned test sites by Peter Goin in his *Nuclear Landscapes* (1991). These latter contribute to the visual archive in interesting ways, fuelling the anti-nuclear sentiment in the process.

Pictures drawn by survivors have appeared in the form of a collection, the Japanese Broadcasting Corporation's *Unforgettable Fire: Pictures Drawn by Atomic Bomb Survivors* (1977). Iconic works such as the Marukis' *The Hiroshima Murals* have been celebrated worldwide and compiled into collections (Dower and Junkerman 1985). Artworks, such as *The Chernobyl Herbarium* (Marder and Tondeur 2016) and the images/specimens of Cornelia Hesse-Honeggar (2008), are also integral to the discourse of planetary harm in the nuclear cultures of the contemporary.

More recently, the graphic novel form, or more accurately, the manga, has been employed as a vehicle to capture and recall the horrors of the bombing in the highly acclaimed autobiographical work of Keiji Nakazawa: *Barefoot Gen* (10 volumes, new English translation 2004–2010) and has come in for critical attention by scholars such as Hillary Chute (2016).

Poetry

Poetry written in the wake of Trinity and the bombings of Japanese cities has been compiled. Powerful poetry about the destruction and personal suffering appeared from poets such as Nanao Sakaki, Shinkichi Takahashi, Tōge Sankichi and others, thereby demonstrating the suppleness of forms in dealing with catastrophic destruction (Bradley 1995). The Hindi poet S. H. Vatsyayan 'Agyeya', also composed 'Hiroshima' where, speaking of the *pika* – the flash of the bomb – Agyeya writes in 'Hiroshima':

On this day, the sun
Appeared – no, not slowly over the horizon
But right in the city square . . .
                    (unpaginated)

Tōge Sankichi's 'The Shadow', about the man vaporized by the bomb and whose shadow was burnt into the concrete, speaks of a 'new Hiroshima' in which the bomb's hypocentre is now a tourist attraction labelled 'Historic A-Bomb Site' (Bradley 22–3). Shinkichi Takahashi in 'Destruction' captures the fragility of all living things, from sparrows to humans (Bradley 297). Kurihara Sadako notes how the A-bomb is explained away as a rightful act for the crimes perpetrated by the Japanese. When we mention 'Hiroshima', writes Sadako, people hear 'Pearl Harbor' or 'Rape of Nanjing' – two events that mark the narrative of Japanese excesses. Terming Hiroshima a 'city of ruthlessness and bitter distrust', Sadako argues that the Japanese will remain forever 'pariahs/scorched by remnant radioactivity' (Bradley 202). Numerous Euro-American poets such as Allen Ginsberg, Edith Sitwell, William Stafford, Mark Sanders, Gary Snyder, Adrienne Rich, Denise Levertov, Richard Wilbur and others have addressed nuclear fear and nuclear harm in different ways. Levertov in 'Watching "Dark Circle"', the celebrated anti-nuclear documentary (1982) criticizes the use of animals in 'simulations' of atomic bomb conditions. This is Levertov's point:

It is
Not a simulation.

As Levertov notes, the pigs are 'real', as is their agony. The 'simulation of hell' is 'hell itself'. She concludes with a half-refrain:

What can redeem them?
                    (Bradley 56)

There is no redemption for the irradiated, because radiation effects are permanent. Indeed, the fear of ingesting anything irradiated is all pervasive. As Sujata Bhatt writes in her poem, 'Wine from Bordeaux', the man who buys two thousand bottles of the wine, will only buy those from 'before Chernobyl'. But 'Chernobyl' is not just *that* location. For the wine-buyer, all Europe is irradiated irreversibly:

He doesn't like
to ingest anything harvested
      in Europe after 1985.
            (Bradley 102)

All Europe has been contaminated, Bhatt suggests.

Plays written about the bombings and the lives of the hibakushas have also since appeared (Goodman 1994).

Films

The Cold War era amplified what Spencer Weart terms 'nuclear fear' in his work of the same title (2012), and popular Hollywood and other cinemas – the many versions of the Japanese *Godzilla* and the celebrityhood of *Hiroshima Mon Amour* (1959) and *The Day After* (1983) being instances – have dwelt on the nuclear 'plot' (Broderick 1996). Documentaries such as *Dark Circle* (1982) and *If You Love this Planet* (1982) but also early animation films such as *One World or None* (1946) drove the discussions around nuclear scare and nuclear threats and have been the subject of extensive studies (Lynch 2012). Advocacy and anti-nuclear propaganda documentaries (*A Hard Rain,* 2007, *Buddha Weeps in Jadugoda* 1999), and the uranium film festival site archives large numbers of these.

Auto/biography

Numerous biographies of the bombmakers – from both sides of the Atlantic – have enabled the creation of the cult of the nuclear scientist (Bernstein, Bird and Sherwin 2005, Thorpe 2006, Bruzzaniti 2016, Segrè 2016, among others). Celebrity events such as the 'inquiry' into Robert Oppenheimer catapulted the already famous nuclear scientist into the realm of the controversial (whose transcript is now available in the United States Atomic Energy Commission's *In the Matter of J. Robert Oppenheimer: Texts of Principal Documents and Letters,* 1971). The 'inquiry' elicited massive responses, mostly in support of Oppenheimer, who became a classic victim of McCarthyism, as Barton Bernstein (1982), David Hecht (2010) and others have observed. As Lloyd Garrison, Oppenheimer's counsel at the hearing, bluntly puts it:

> There is more than Dr. Oppenheimer on trial in this room. I use the word 'trial' advisedly. The Government of the United States is here on trial also . . . There is anxiety abroad that these security procedures will be applied artificially, rigidly, like some kind of machine that will result in the destruction of men of great gifts and of great usefulness to the country . . . America must not devour her own children.
>
> (US Atomic Energy Commission 990)

Such events also injected elements such as the fear of the 'communist bomb' into the discourse of nuclearity.

Autobiographical accounts, reflections and statements about the bomb, some written in the wake of the Hiroshima–Nagasaki destruction by members of organizations such as the Federation of Atomic Scientists, the obituaries of famous nuclear scientists and observations by scientists on other scientists are also integral to nuclear cultures (Oppenheimer 1979, Bethe 1997). Campaigns and advocacy movements, manifestoes, petitions and public letters, many by the world's top nuclear scientists, seeking curbs, international control and cautioning the world that a nuclear arms race would destroy the world (Masters and Way 1946, Bohr 1950, Russell-Einstein 1955) also add a different dimension to nuclear cultures. The Masters-Way collection, *One World or None*, brought together Einstein, Bethe, Oppenheimer and other globally recognized scientists to speak of the future of a nuclearized world and may be read as one of the first advocacy texts that highlighted planetary precarity.

Naturally, this book does not seek to cover all aspects of nuclear cultures. For instance, it eschews histories of institutions such as the Atomic Energy Commission, movements such as the scientists' movement in America and histories of nuclear energy and discoveries, documented in detail by Alice Kimball Smith in *A Peril and a Hope: The Scientists' Movement in America 1945-47* (1971). It also steers away from the vast memorial cultures around Hiroshima and Nagasaki, already explored in considerable and precise detail by other, and more accomplished, commentators with access to material in Japanese and other languages (Yoneyama 1999). It also does not address the *popularity* of nuclear power, or the campaigns for its enhanced use. The political cultures of the bomb, nuclear policy, the 'North-South' divide in nuclearization are also not the subject of this book, although the first – political cultures – may be inferred in many of the texts studied here.

The 'Work' of Nuclear Cultures

As this book sees it, the literary-cultural texts that make up the nuclear cultures of the contemporary, even when sampled selectively, present an unrelenting prognosis of nuclear harm to the entire planet, even as they hark back to the past in their attempts at memorializing. That is, nuclear cultures explore ways for the memorialization, even aestheticization, of these events but also seek to strike the cautionary note or paint a hor-rific prognosis of a nuclearized world. Works like that of the Marukis link multiple holocausts, from Hiroshima to Auschwitz – leading to the creation of 'global memory cultures', as Ray Zwigenberg argues (2014. Also Wake 2021). Initiatives such as Norman Cousins' controversial 'Moral Adoption' programme and the '*Hibakusha* Maidens' event of 1955 also mobilized sentiment although, as Lifton and Mitchell note in their justly famous study, *Hiroshima in America: A Half Century of Denial* (1995), it attracted its share of criticism especially from the US government (254–5).

There is considerable diversity within the genres and forms of nuclear cultures, as the previous section indicates in its short inventory of texts. In their efforts to capture nuclear harm, these forms and genres have evolved their own discourses, adapted other discourses or merged mul-tiple discourses. Hence, cataclysmic events like the nuclear bombing that demand specific aesthetic strategies find literary and visual texts developing new modes of representation. To take one example for now, Ibuse's *Black Rain*, described as a 'documentary novel' by its translator (8), also employed metaphors to speak of the sights that unfolded in Hiroshima on that day:

> Along the railroad tracks there stretched a long train of refugees like a trail of ants, or like the pilgrims – it occurred to me – who were said to have lined the approaches to the shrines of Kumano in olden days. Seen in the distance, the hill in the park was like nothing so much as a great pale bun with ants swarming all over it.
>
> (47)

Or take Furukama in his Fukushima novel, *Horses, Horses, in the End the Light Remains Pure*, where the epithet of the 'land of the rising sun' is inverted after the disaster:

> How does one sing praises to this national land? Especially now, given that there is a second sun in the nuclear core?
>
> (65)

In Yoko Tawada's *The Last Children of Tokyo* (2018), set in a post-apocalyptic world of strange mutations, children are born with brittle bones and soft teeth and the elderly are youthful ('grown-ups can live if children die, but if grown-ups die children can't live', says the child Mumei in the novel, 32). The men experience menopausal symptoms. Tawada does not give us temporal markers to help us locate the events in any definite future, but seems to indicate that generations to come will suffer from bizarre illnesses. The persistence of contamination, Tawada summarizes as follows:

> While calves drink their mother's milk, baby birds eat the worms their parents bring them. But worms live in the earth, so when the earth is contaminated, the contamination gets concentrated in the worms . . .
>
> (20)

The bomb-injured bodies demanded a toxic somatography when writers wished to document them. The Marukis employed Japanese hell-painting traditions to depict disaster. Then there are the Peter Goin photographs of the landscapes of nuclear test sites, and the atomic-industrial sublime of the laboratories and factories in written histories of the making of the atomic-bomb. But it is not only in the bomb's explosive power and the quantum of destruction that the artists and authors (or this book) are interested in. Equally important for nuclear culture is the cult of the nuclear scientist and the 'making' of the world's most terrifying weaponry.

Constructions of the personae of the bombmakers represent their charisma, scientific authority and instantiate in many cases a nuclear patriotism. In explicitly anti-nuclear texts, the symbols, advocacy rhetoric, the non-human and the concrete structures that make up and inhabit the landscapes of uranium mining, nuclear testing and nuclear disaster, and gesture at planetary ruination are constitutive of nuclear cultures.

The work of nuclear cultures is the prognosis of planetary precarity, and authors, artists and commentators seek to find appropriate modes of communicating this prognosis. Taken together, the archive of materials – fiction, photographs, drawings, memoirs, histories, biographies, films and comics – capture in Debjani Ganguly's evocative phrase, 'global histories of irradiated violence' (437). They indicate that the notion of 'isolates' and regions safe from nuclear harm is a myth (deLoughrey 2012). In the words of Hideo Furukawa commenting on the creation of zones of danger/exclusion after the disaster: 'can one truly escape by leaving the prefecture?', suggesting that there is no place to escape to, not from the radioactive fallout (25).

Planetary precarity, as the texts in this admittedly programmatic study indicate, is primarily the consequence of the nuclearization of the world, from the Trinity test in 1945 through the Cold War's arms race and escalated the production of nuclear weapons to the disasters at Three Mile Island, Chernobyl and Fukushima (to cite only the major ones). Planetary precarity is an expansion of 'ecoprecarity'. If ' "ecoprecarity" is at once about the precarious lives *humans* lead in the event of ecological disaster – witness Katrina, Fukushima, the tsunami and, of course, Bhopal and Chernobyl – and also about the environment itself which is rendered precarious due to human intervention' (Nayar 2019: 7, emphasis in original), planetary precarity underscores that there is *no* region, species or structure that is not always already nuclearized for an infinite duration of time, given the half-lives of polonium, uranium and cesium. Jean-Luc Nancy writes:

> Nuclear catastrophe – all differences military or civilian kept in mind – remains the one potentially irremediable catastrophe, whose effects spread through generations, through the layers of the earth; these effects have an impact on all living things and on the large-scale organization of energy production, hence on consumption as well.
> (3. Nancy would use the term 'irremediable' throughout the work, to speak of human civilization)

Nancy argues that all catastrophes hereafter are nuclear in nature:

> the spread or proliferation of repercussions from every kind of disaster hereafter will bear the mark of that paradigm represented by nuclear risk.

(3)

The human, non-human, geological, atmospheric have been irradiated nuclear 'subjects' since 1945. There are no definite numbers of the affected for, as Alexander Borovoi puts it in *My Chernobyl*: 'millions of human lives were sucked into the whirlpool of this tragedy, and every life was refracted and distorted' (2017: 10). And there is no timeline for the nuclear harm to be reduced in intensity.

As extinction studies (Colebrooke 2014, Rose et al. 2017), thought-experiments (Weismann 2007) and climate trauma studies (Kaplan 2016) have indicated, texts that carry a prognosis of apocalypse encode a 'Pretraumatic Stress Syndrome' (Kaplan 1) and are often elegiac – perhaps an elegy in advance, for the human race but also all life forms? – in tone (Heise 2008). There are, however, more than the elegiac and the apocalyptic modes of communicating the sense of an ending – whether planetary or species – as this book captures in its reading of a diverse set of texts from all over the planet. The larger aim of nuclear cultures and of this book is to demonstrate a set of discourses, aesthetics, representational strategies that communicate this precarity. The examples, which are obviously a miniscule sample of the nuclear archive, serve to show connections *across* discourses and narrative strategies, perhaps in the process being untrue to the aesthetic or discursive features of specific genres.

The book has four core chapters.

The second chapter examines two questions: how does a nuclear subject emerge in literary-cultural texts? What *is* the figure of the nuclear subject? As response to these interconnected questions, it forwards the answer that the irradiated body is the locus of nuclear subjectivity. The emerging nuclear subject appears in a genre I term 'toxic somatography'. Other texts construct the nuclear subject through an emphasis on the communitarian connections of the irradiated body. Finally, the irradiated body is also the ground upon which a reconstruction of the subject begins. The chapter proposes that the everyday lives of people in the wake of nuclear bombings, disaster or testing have been rendered strange and unfamiliar ever since their settings, ecosystems and habitats are irradiated by something toxic and invisible. Living in the familiar-yet-strange spaces and telescoping temporality generates the experience of the 'nuclear uncanny'.

Chapter 3 examines a different version of the nuclear subject: the atomic scientists and bombmakers, and the persona of the bombmakers as it emerges from the biographies. My focus is not the self-fashioning by the scientists themselves – although there is that too – but the *biographical* construction of the atomic scientist persona. This focus on the biographical construction of the scientific persona enables us to locate the atomic scientist within three crucial contexts – the Second World War, the Hiroshima–Nagasaki bombings, and the subsequent campaign for delimitation and control of the nuclear arsenal. The biographical construction of the atomic scientist persona reveals multiple layers and nuances of what counted as a 'scientific self' and, frequently, the origins of this self in the scientist's childhood, modes of working and the milieu.

Chapter 4 is a reading of the aesthetics of nuclear disaster as employed by different kinds of texts. The nuclear sublime is an irradiated aesthetics which captures a state of permanent crisis. The nuclear sublime is more than the aesthetics of an event or its descriptions in visual and verbal texts: it is a horror-aesthetics of a continuing crisis and permanent damage across the world. How exactly such an aesthetics operates is the subject of this chapter. It does not privilege the sublime's distanced view but tempers it with the close, and haunting, encounters with mass death, the spectacle of large-scale injuries and devastation of habitations and neighbourhoods. The chapter also unpacks an atomic sublime in the making and operations of the atomic-industrial complex.

Chapter 5 examines the discourses and representational strategies in the anti-nuke movement, through a reading of texts from different places, from Navajo nation art to Maori fiction to documentary films about India's uranium mining. I examine the mobilization of anti-nuclear sentiments, discourses and texts from campaigns (including those against nuclear waste, uranium mining and reactors) foregrounding all the while, the risks and threats posed by nuclear power. It studies the rhetoric of place in such campaigns, the rise of anti-nuclear cosmopolitanism and the nuclearized non-human. It concludes by proposing two cosmograms for the nuclear age – the radiation hazard symbol and the Doomsday Clock.

Note

1 https://mitpress.mit.edu/books/series/radium-age

2 The Nuclear Subject

George Weller, who describes himself as 'the first Allied observer to reach Nagasaki since the surrender' describes his tour of the ruined city. He recounts the sights of devastation:

> Several children, some burned and others unburned but with patches of hair falling out, are sitting with their mothers. Yesterday Japanese photographers took many pictures of them. About one in five is heavily bandaged, but none are showing signs of pain.
>
> Some adults are in pain as they lie on mats. They moan softly. One woman caring for her husband shows eyes dim with tears. It is a piteous scene and your official guide studies your face covertly to see if you are moved.
>
> (unpaginated)

Soon after, Weller provides an autopsy account from Dr. Yosisada Nakashima:

> Interior symptoms of the first class revealed in the post-mortems seem to show the intestines choked with blood, which Nakashima thinks occurs a few hours before death. The stomach is also choked with blood, and also mesenterium. Blood spots appear in the bone marrow, and subarachnoid oval blood patches appear on the brain which, however, is not affected. Upgoing parts of the intestines have little blood, but the congestion is mainly in downgoing passages. The duodenum is drained of blood, but the liver, kidney and pancreas remain the same. The spleen is hard but normal, though the urine shows increased blood.
>
> (unpaginated)

In the immediate aftermath of the Buffalo and Antler tests in Australia, John Moroney, Secretary of the Atomic Weapons Tests Safety Committee, sent the following cable to H.J. Brown, Controller of the Weapons Research Establishment at Woomera:

DOI: 10.4324/9781003254294-2

(1) Did a mystery disease of epidemic proportions a few months ago result in a number of deaths amongst Aboriginal children at the Ernabella Mission Station in South Australia? (2) Is it a fact that in certain quarters the deaths of these children were attributed to the effects of radioactive fallout from bomb tests? (3) If he has not already done so, will he have this report investigated immediately and make information available as soon as it comes to hand?

Brown responded:

a disease of epidemic proportions did occur at Ernabella during March to June 1957, resulting in the deaths of 20 children and 2 adults.

(cited in Tynan, unpaginated)

The Weller account records the survivor bodies in post-atomic bomb Japan. Tynan records the effects of nuclear testing in Australia, a decade *after* the atomic bombing of Japan. Both construct the 'nuclear subject' by focusing on the irradiated body *and* the irradiated community.

The nuclear subject emerges in multiple texts from different geocultural zones and across time spans. Admittedly, in the case of the *hibakushas*, an institutionalization of the nuclear subject has occurred, through the medical, legal and activist discourses that have informed the practice of the testimony genre. Lisa Yoneyama argues:

Because it measured damage by calculating spatial and temporal proximity to the location and the moment of explosion, classifying individual survivors accordingly, survivors' accounts also tended to be saturated with exact figures and scientific terms. References to precise and detailed data on the number of casualties, the temperature of heat rays, the strength of the atomic blast, and the height of the bomb's explosion helped fashion survivors' accounts, translating the catastrophe into measurable and calculable damages.

(94)[1]

In what follows, I sidestep the question of memorialization and testimony and instead focus on other aspects of the nuclear subject, examining such a subject as s/he emerges in events and discourses from around the world.

While one manifestation of the nuclear subject may be found in the survivor accounts, memoirs and images (drawings, photographs, graphic texts) from Hiroshima and Nagasaki, another may be found in texts (poetry, auto/biographies, fiction) from Nevada, the Marshall Islands and Australia, by/of victims-at-a-distance of nuclear tests and nuclear disasters such as Chernobyl (oral histories, fiction). These narratives construct subjects living in a state that Spencer R. Weart termed 'nuclear fear' (2012), and occasionally staring at near/extinction. The rise of the nuclear subject, its contours, is the focus of this chapter, which makes three arguments.

The irradiated body as the locus of nuclear subjectivity and subject is constructed in these narratives as a 'material witness' (Schuppli 2020). To argue for the human person's body as a material witness is *not* to reduce it or treat it as a non-living 'thing', but rather to emphasize how the body of an individual and populations serve as evidentiary texts for the effects of political violence such as the bombing of Japan, the evacuation of the Marshall Islanders to make way for nuclear testing, or the devaluing of the lives of the Aborigines of Australia or the 'downwinders' in Nevada in the process of testing. The irradiated body of the nuclear subject is also visible through a genre I term toxic somatography. Other texts, however, also construct the nuclear subject through an emphasis on the communitarian connections and linkages of the irradiated body. Finally, the irradiated body is also the ground upon which a tentative reconstruction of the subject begins.

The chapter also proposes, following the work of Annelise Roberts (2021) and Saint-Amour (2000), that the everyday lives of people in the wake of nuclear bombings or testing have been rendered strange and unfamiliar ever since their settings, ecosystems and habitats are irradiated by something toxic and invisible. Living in the familiar-yet-strange spaces and telescoping temporality generates the experience of the 'nuclear uncanny'. Here too the irradiated body is the locus of a nuclear subject.

The chapter's bringing together of narratives and memories from different locations and contexts is intended to drive home the point that the nuclear subject, from Utah to Hiroshima, Rongelap to Chernobyl, is forged in the crucible of omnicidal science and technology (nuclear) wherever they are located. It is also meant to capture the planetary nature of nuclear vulnerability. Nuclear vulnerability and conditions of precarity – the heart of the nuclear subject – are less about nationalist discourses and identities than about a global collective aware of the histories as diverse as that of the 'downwinders' in Utah, the victims in Hiroshima–Nagasaki, the returning populations of Prypjat and the continuing sufferers of Bikini Atoll. Geographical, demographic and cultural 'isolates', as Elizabeth deLoughrey (2012) argues, is a just a myth, whether we are speaking of tiny islands or the expansive continent, when we think of nuclear subjects.

It should further be underscored that the Japanese – the first victims of the atomic bomb – were dehumanized by Euro-American cultural narratives throughout the war and hence were nuclear subjects of an entirely different order:

> Alongside popular fiction and documentary films and publications, relevant United States industry periodicals, particularly in the later years, attest to a mindset in which the Japanese enemy had been

sufficiently dehumanised as to be depicted and narrated as resembling human-insect hybrids for whom 'pest-control' was the 'solution'.

(Broinowski 94)

In other words, the Japanese had already been cast as non-human, or semi-human nuclear subjects in time for the atomic bomb. In what is surely a twisted irony, such subjects were also constructed in the downwinders of Utah, the natives of Marshall Islands and the Aborigines of the Australian outback.

The Irradiated Body

The irradiated body in nuclear texts is a material witness. I adapt the term 'material witness' from Susan Schuppli's work. Schuppli's focus is the non-human:

> *Material witnesses* are nonhuman entities and machinic ecologies that archive their complex interactions with the world, producing ontological transformations and informatic dispositions that can be forensically decoded and reassembled back into a history. *Material witnesses* operate as double agents: harboring direct evidence of events as well as providing circumstantial evidence of the interlocutory methods and epistemic frameworks whereby such matter comes to be consequential. *Material witness* is, in effect, a Möbius-like concept that continually twists between divulging "evidence of the event" and exposing the "event of evidence."
>
> (3, emphasis in original)

Indeed, the employment of the term signals the *terra nullius* subtext to the discourse of nuclear testing/bombing precisely because the inhabitants of Nevada, Hiroshima or the Marshall Islands were deemed non-persons, and the land was 'empty' (I address the nuclear *terra nullius* in Chapter 5). But the invocation of bodies as material witnesses also serves the purpose of examining the corporeal-centric discourse of nuclear effects in survivor and other texts.

It is of course important to note that the human body is a material witness alongside buildings, roads, bridges, the non-human and other such objects. For Schuppli, these latter are 'technical witnesses', 'that which can offer material evidence of events but cannot testify of its own accord' (10). Schuppli's attempt, as she puts it, is to examine the 'expressive technicity of matter' (11). While Schuppli is correct in making a distinction between the human witness and the technical witness, it is also undeniable that the human person who does not 'testify of [her/his] own accord' can also be made to tell the story, for, as Jacques Derrida pointed out in his work on the testimony, very often the witness who cannot tell his story offers himself *as* the story (2000). Besides, as we shall see, the human person who does not testify of her/his own accord, or even when no longer living, is the source of a testimony in the form of the autopsy, the postmortem and the eyewitness account. In other words, instead of separating the witnessing of/by material objects from the witnessing of/by humans, I treat the human person as *matter* that witnesses and experiences large-scale suffering and destruction.

The human as material witness finds its strongest expression in a toxic somatography and the medicalized somatography of survivor narratives.

Toxic Somatography and Atomic Trauma

Central to the construction of the nuclear subject is the image – visual, verbal – of the scarred, burned, injured body. Bodies *visibly*, that is *materially*, marked by radiation constitute the bulk of narratives of Hiroshima and Nagasaki, described through the mode of documentary realism. All survivor and eyewitness narratives are examples of a toxic somatography – a term I adapt from Thomas Couser's 'somatography' in his work on the disability memoir, to signify writing about living in and as a particular 'anomalous' body (Couser 2012: 2). In other words, radiation, which is invisible, is perceived in the form of gruesome material effects: on the human body. Toxic somatography is the account of these damaged bodies where the toxin itself is intangible and yet palpable. The toxic somatography of atomic trauma captures a constant play of visible/ invisible, inside/outside when speaking of the injured and the dead in atomic disasters.

In Tomoko Konishi's drawings and narrative of witnessing in the collection *Unforgettable Fire: Pictures Drawn by Atomic Bomb Survivors* (1977), he gives us this: a girl squatting on the road with what appears to be a bit of cloth hanging down from her back. The illustration is annotated in Japanese, with arrows pointing to various features in the image, and the English text follows beneath the image:

> Her back was completely burned and her skin peeled off and was hanging down from her hips.
>
> (13)

In other illustrations, we are shown people with 'the flesh . . . scooped out and bleeding profusely' (13) and 'skin of both . . . hands . . . hanging loose as if it were rubber gloves' (13).

Tada Makiko describes her situation within a few minutes of the bomb:

> my body was burning pale . . . I washed my face . . . but I just couldn't get rid of something sticky, and when I touched and pulled it with my hand, the thin skin-like thing stuck and wouldn't come off . . . after bleeding badly, two bloody lumps passed and I fell, my eyes reeling.

When I came to after a while, I pulled a root of the brushwood and poked the lumps with it, and it was as if I were poking a liver. I began to feel a little better while lying down, but, according to my child, I became weird around then . . .

(Kyoko and Selden 174–5)

Dr. Tabuchi tells Hachiya of the sights he witnessed:

Hundreds of injured people who were trying to escape to the hills passed our house. The sight of them was almost unbearable. Their faces and hands were burnt and swollen; and great sheets of skin had peeled away from their tissues to hang down like rags on a scarecrow. They moved like a line of ants. All through the night, they went past our house, but this morning they had stopped. I found them lying on both sides of the road so thick that it was impossible to pass without stepping on them.

(Hachiya)

Takashi Nagai, the physician injured in Nagasaki and who became iconized as a saint for working despite his wounds, in *The Bells of Nagasaki* (1984), describes how, as he went about treating the other injured, he experienced dizzy spells, his own injuries opened up, etc. (44–5), before reflecting on the possible effects of the bomb (52–6). Nagai is alert to his own body as specimen, constantly reflecting on his nausea (61) even as he treats the patients of similar symptoms, and its possible reasons (gamma rays, 61) and the extent of what he calls 'atomic sickness' (62, 63, 83–4 and elsewhere). He comments on the nature of skin burns and computes the damage in terms of distance from the hypocenter of the blast (65).

Throughout the first two volumes of Nakazawa's *Barefoot Gen* (2004), we have grotesque irradiated bodies. Flesh and innards fall from the bones of the living (both human and non-human), the faces have pus-and-maggot-filled injuries, the river is full of bloated corpses that burst open from the gases inside, the landscape is full of skulls and bones and mass graves (I: 254, 256; II: 34–5, 51, 75, 87–8, 104). Nakaoke finds himself sheathed in flies (57) from the dead. Bodies burst out of the funeral pyres (171–3). As the medics and rescue workers seek to help the burnt individuals by hauling them into the trucks, the flesh slides off the bones (II: 83, 94). It is significant that Nakazawa repeats the images of the death of his father, sister and brother in the bombing (IV: 5, V: 8, 144, 211, 245, VI: 198, 199, VII: 190 and elsewhere), and the bomb victims staggering off with their severely damaged bodies, across multiple volumes (IV: 258, V: 7) – all of which underlines a *continuity* of trauma at the memories of burning bodies, dying family members and ruined homes. Indeed, Nakazawa titles volume V, 'the never-ending war' precisely to capture continuing traumas and life *after* the bomb (which is the title of volume III).

John Hersey cites another eyewitness' recall of Hiroshima:

> he met hundreds and hundreds who were fleeing, and every one of
> them seemed to be hurt in some way. The eyebrows of some were
> burned off and skin hung from their faces and hands. Others, because
> of pain, held their arms up as if carrying something in both hands.
> Some were vomiting as they walked. Many were naked or in shreds of
> clothing. On some undressed bodies, the burns had made patterns –
> of undershirt straps and suspenders and, on the skin of some women
> (since white repelled the heat from the bomb and dark clothes
> absorbed it and conducted it to the skin), the shapes of flowers they
> had had on their kimonos.
>
> (unpaginated)

Life magazine's special report on the bombing, published on 20 August
1945, quoted a Japanese eyewitness account:

> All around I found dead and wounded . . . bloated . . . burned with a
> huge blister . . . all green vegetation perished.
>
> ('War's Ending' 26)

The human body in each case is a material witness to the event of the
bombing, the bomb's locus, so to speak. The body as material witness has
two components. First, the total *dematerialization* of the human form,
where eyewitnesses report people being vaporized with the blast, and
second, the debodiment of the materiality of the person.

In Wilfred Burchett's report, titled 'The Atomic Plague', September
1945, he writes:

> Of thousands of others, nearer the centre of the explosion, there was
> no trace. They vanished. The theory in Hiroshima is that the atomic
> heat was so great that they burned instantly to ashes – except that
> there were no ashes.
>
> (unpaginated)

Some reported how Hiroshima itself seemed to have vanished:

> Hiroshima had disappeared . . . that experience looking down and
> finding nothing left of Hiroshima was so shocking that I simply can't
> express what I felt. . . .Hiroshima didn't exist-that was mainly what I
> saw – Hiroshima just didn't exist.
>
> (cited in Kyoko and Selden xx)

In *Unforgettable Fire*, Konishi describes his father's disappearance (13).
The dematerialization of persons and places is oddly and tragically

accompanied by a strange materialization in the form of shadows and imprints of the persons who are no longer there. In the famous image from Hiroshima, the shadow of the persons who leaned against the stone pillar or sat on stone steps were imprinted on the stone. The caption to this image, included in the *Effects of the Atomic Bombs: Report of the British Mission to Japan* (1946), says:

> the polish [on the granite] remains only where shielded by (a) a man seated on the steps, (b) a man leaning against the corner of the plinth adjoining the steps and (c) in the "shadows" of the plinth mouldings.
> (Photograph # 21, Effects of the Atomic Bombs)

This is the dematerialization and re-materialization of the human person, and an image of annihilation itself. As Akira Mizuta Lippit puts it:

> [these are] frames of an annihilating image, an image of annihilation, but also the annihilation of images, no one survives, nothing remains: "It made angels out of everybody."
>
> (95)

He adds:

> At Hiroshima and Nagasaki, two views of invisibility – absolute visibility and total transparency – unfolded under the brilliant force of the atomic blasts. Instantly penetrated by the massive force of radiation, the *hibakusha* were seared into the environment with the photographic certainty of having been there. In the aftermath of the bombings, the remaining bodies absorbed and *were absorbed by* the invisible radiation. These bodies vanished slowly until there was nothing left but their negatives.
>
> (95, emphasis in original)

The person's materiality is erased and re-materialized in the form of material evidence – the shadow on the granite – simultaneously. The granite then is a host to a material witness (the human's shadow) of/to matter that is no longer present (the human body). Numerous references to this startling de- and re-materialization occur in literary texts.

The poet Sankichi in 'The Shadow' would speak of 'someone's trunk' burned into the granite, and for survivors 'this shadow/etched in tragic memory' will never fade (Bradley 22–3). William Dickey references both this burnt-into-the-stone shadow and Hersey's *Hiroshima* in his poem 'Armageddon' (Bradley 137–9), clearly demonstrating the widespread circulation of this material photograph of someone who was no longer matter. Nanao Sakaki in 'Memorandum' writes of searching for a missing friend in Hiroshima one year after the bombing.

As a substitute for him
I found a shadow man.
The atomic ray instantly
disintegrated his whole body,
All – but shadow alive on concrete steps.
(Bradley218-19)

Now, if Lippit argues that 'total materialization and total dematerialization institute the same crisis in visuality' (83), the argument can be extended to the verbal text, where the account of dematerialization of the human form demands a re-materialization in the form of the archive of corporeal suffering and accounts of vanished humans. Even when the human is no longer visible as *matter*, s/he remained materialized in the shadow imprinted on stones.

While the visibly marked body of atomic trauma is the obvious departure point for the toxic somatographic narrative, a second modality of the survivor/eyewitness account is also discernible in the theme of bioaccumulation. Bioaccumulation refers to 'the processes by which toxic substances, industrial waste, and man-made chemical compounds gradually accumulate in living tissue' (Radomska and Åsberg 60). Bioaccumulation is the source of the marked body's toxic embodiment, 'a condition where differentially situated human and non-human bodies, land- and waterscapes are immersed in the naturalcultural intra- and interactions with toxicity' (60). Numerous instances of bioaccumulation may be found in the nuclear narratives.

Wilfred Burchett was one of the first to point out that

people who, when the bomb fell, suffered absolutely no injuries, but now are dying from the uncanny after-effects. For no apparent reason their health began to fail.

(unpaginated)

Then, and only then, would the body become marked:

They lost appetite. Their hair fell out. Bluish spots appeared on their bodies. And the bleeding began from the ears, nose and mouth.

(Burchett, unpaginated)

The bodies experienced toxic bioaccumulation, coming from sources as diverse as the soil and the water in Hiroshima:

They found that the water had been poisoned by chemical reaction. Even today every drop of water consumed in Hiroshima comes from other cities. The people of Hiroshima are still afraid.

(Burchett, unpaginated)

The bodies of those not immediately killed began to implode as a result of the radiation. 'Everybody's resistance was reduced by the radiation . . . ', writes Hiroko Takanashi in the collection *Hiroshima: Survivors' Testimonies* (20). Conditions and symptoms – such as bleeding from the lungs – emerged over time. 'I have constantly been anxious about atomic bomb disease', says Akiko Mizue in *Hiroshima: Survivors' Testimonies* (37). Thus, those with no visible sign of injury were also dying on the inside although they did not know it then, as George Weller notes:

> These people squatting before me had run around, salvaging, unworried, believing they were safe because they were unburned. Then, carelessly, they scratched a finger on broken glass, or bit off a hangnail. And they bled. And they bled.

From four decades later, but with the same theme of toxic somatography, Svetlana Alexievich observes:

> [they were] people whose world was Chernobyl; for them, everything inside and out was poisoned, not just the land and the water.
>
> (27)

And elsewhere:

> Our eyes, ears and fingers were no longer any help, they could serve no purpose, because radiation is invisible, with no smell or sound. It is incorporeal.
>
> (28)

One resident tells her:

> It may be poisoned with radiation, but this is my home.
>
> (48)

But the 'incorporeal' nature of radiation poisoning does have a material witness: and Alexievich documents oral accounts from men and women who developed leukaemia after coming back from Chernobyl.

Nuclear subjectivity emerges in this interplay between the harm and trauma etched on to the bodies of victims through the blast, and the implosion of the body exposed to invisible radiation over a period of time. That is, the body is inscribed internally with the radiation, and it implodes. We see this implosion trope in Shinkichi Takahashi's 'Explosion' where the speaker describes how, after the bomb, s/he breathes in radioactive air:

> A billion years
> And I'll be shrunk to half,
>
> (Bradley 298)

The nuclear subject is one whose subjectivity is grounded in the terror of the invisible, which the subject feels, and believes is internal – as deep as the marrow – to the body now, as much as it is the fear of the bomb. In this, toxic somatography is a nuclear equivalent of a feature of the Gothic:

> Terror . . . is the *frisson* that is provoked by the invisible, by what lurks unseen in the dark. Therefore, I reasoned that the texts would likely provoke horror in response to visible disabilities like bodily deformity, and terror in response to invisible disabilities like sense disabilities and infection.
>
> (Anolik 8)

The sickening, imploding body is an index of invisible bioaccumulation: it *embodies* in its very materiality, an all-pervasive but invisible radiation. The terror originates in the corporeal appearance of something invisible.

We see this in accounts of the non-military nuclear testing in the Marshall Islands, including Bikini Atoll. In fact, nuclear testing left material evidence on human bodies and elsewhere, on islands such as Rongelap. Lemoyo Abon recalls the tests at Rongelap:

> After noon, something powdery fell from the sky. Only later were we told it was fallout. With Roko and several cousins, I went to our village on the end of Rongelap island to gather some sprouted coconuts. One cousin climbed the coconut tree and got something in her eyes, so we sent another one up. The same thing happened to her. When we went home – ours was the main village on Rongelap – it was raining. We saw something on the leaves, something yellow. Our parents asked, 'What's happened to your hair?' It looked like we'd rubbed soap powder in it.
>
> That night we couldn't sleep, our skin itched so much. On our feet were burns, as if from hot water. Our hair fell out.
>
> (cited in Dibblin, unpaginated)

Others recall:

> Several hours later the powder began to fall on Rongelap. We saw four planes fly overhead, and we thought perhaps the planes had dropped this powder, which covered our island and stuck to our bodies.
>
> (cited in Dibblin, unpaginated)

Jane Dibblin, in her history of the tests in Bikini, writes:

> The pale powder continued to fall until late afternoon, by which time it was about one and a half inches deep. Later it emerged that it was

in fact particles of lime (calcium oxide) formed when Bikini's coral reef (a formation of calcium carbonate) melted in the intense heat of the bomb and was sucked up and scattered for miles. The exact dose of radiation received by the islanders was never measured, but it was estimated that people on Utrik received 14 rem (140 msv) and those on Rongelap 175 rem (1,750 msv). The International Commission on Radiological Protection (ICRP) now recommends that a maximum permissible total body dose to a member of the general public be 0.5 rem a year.

(unpaginated)

In Jessie Lennon's autobiography – the first Aboriginal writing on the nuclear testing in Australia – she records her cancer from the bomb (128). As for the other members of the community, she writes, 'for a long time we've been talking Maralinga' (130). The insistence on invisibility and continuity of toxification in all the survivor-witness accounts transforms the corporeal into material witness: one can only discern the toxicity in terms of eroding human bodies, often over an extended period of time.

Specimen, Spectacle and the Medicalized Somatography

Ōishi Matashichi, survivor from the *Lucky Dragon* incident, dispassionately records his cell counts and includes a photograph of him receiving a blood transfusion (2011). In the process, he serves up a medical case study with visual evidence of his condition, diagnostic procedure and prognosis.

In Keiji Nakazawa's *Barefoot Gen*, the painter Seiji, covered in radiation burns and treated as an outcast by his own family, decides that he has had enough of hiding away. He announces to the shocked boys:

> I'll make a spectacle of myself. I'll make sure they never forget these burns and scars . . . I can't let people forget what the bomb has done.
>
> (III: 134)

With the help of Gen and Ryuta, the two boys, he undertakes a tour of the town, having unravelled his bandages (which cover his entire body) and exhibiting his injuries, sores and burns. Nakazawa draws Seiji being carted around the town, shocking people, who scream 'monster' (III: 135). Seiji then becomes a specimen and a spectacle, the material evidence writ large on the effects of the atomic bomb. In Betsuyaku Minoru's play, *The Elephant*, the 'invalid' – the character is known as just that – is one who has been exhibiting his keloid scars to the public in the town square. He recalls an incident where a girl arrived and was fascinated by his scars.

> That kid, she said she wanted to touch the keloid on my back . . .
> she touched me for just an instant and snapped her hand back fast
> as you please.

(Goodman 201)

The epidermalization – to borrow a Frantz Fanon term – of radiation
sickness is best exemplified in Minoru's statement where after a time,
even the scars are ignored by the people. The 'invalid' complains:

> Every day I'd go out into the town, and I'd spread my mat on the
> ground and just stand there daydreaming. The bastards didn't even
> notice. The naked keloid man's sold out his supply of charm and
> moved on to pathos, that's what they said.

(Goodman 202)

Toxic somatographies that construct the irradiated body as a material
witness to the blast and the radiation also offer another set of representational strategies. The individual victims – living and dead – are both
specimen and spectacle in the narratives.

The injured and the dead serve the medical practitioners the first
specimens of atomic trauma injury and radiation sickness. Beyond the
accounts and reports such as *Effects of the Atomic Bombs: Report of the
British Mission to Japan* (1946) or *The Report of the Joint Commission
for the Investigation of the Effects of the Atomic Bomb in Japan* (USA's
Atomic Energy Commission, 1951) which were of course focused on the
victims as specimens, we have memoirs by physicians which use the same
rhetoric of analysis and examination and medical curiosity. (Although
this was 'research without treatment', notes Susan Lindee.)

The victim's body – dead or alive – reveals, as a material witness, the
insidious effects of radiation poisoning. Their organs, having suffered
bioaccumulation on a massive scale, collapse. This collapse is revealed in
the case of the dead through the autopsy, which occupies a central role in
physician memoirs, such as Michihiko Hachiya's, or through the medical
examination in the case of living victims. Hachiya himself was injured
and severely burned in the bombing and was barely recovered when he
began attending to the hospital work again. His entry of 31 August 1945
consists in the main of medical tests and autopsy accounts.

> Going to the Out-Patient Department, I found patients lined up to get
> their blood examined. Dr. Hanaoka and the medical students were
> busy examining blood smears, and on the table I noticed a reagent
> bottle marked "for platelet examination."
>
> "You are making platelet counts," I observed.
>
> "Yes," answered Dr. Hanaoka, "we are, but so many slides are
> totally devoid of platelets there is nothing to count."

(146)

Michihiko then offers a comment:

> Dr. Hanaoka's remarks recalled the autopsy cases. Failure of the blood to clot might well have been because of a decrease in blood platelets. I could hardly wait to confront Dr. Tamagawa with my suspicions.
>
> "Is that so!" he exclaimed. "Well! That explains everything. Yes, indeed! That's why blood hasn't clotted even after seven hours!"
>
> A cloud seemed to lift above Dr. Tamagawa and left him communicative and expansive by contrast with his brusk, noncommunicative behavior earlier. It was as though my comments provided him with the key to a puzzle.
>
> (146)

The 'puzzle' here is the impossibility of understanding the true nature of the material witness: of frameworks to interpret the data from the blood work, or the human body itself. But the puzzle is also a contributory factor to the construction of the doctor's authority and subjectivity. Incidentally, such a 'puzzle' was not restricted to the Hiroshima doctors alone but was the state of affairs in the USA many years after the bombing, as Sarah Alisabeth Fox notes in her study of the Utah atomic tests:

> since scientists knew relatively little about the by-products resulting from the tests, the way they traveled through the environment, or their potential effects on humans, many toxic radionuclides were never monitored at all.
>
> (10)

Soon after this, Michihiko would document an autopsy result:

> He paused to show me petechiae visible in organs he was removing from a patient and expressed the opinion that a decrease in blood platelets was responsible for the development of petechiae. His cases and the case autopsied by Dr. Katsube showed the same changes . . .
>
> Miss Kobayashi, who died on the twenty-sixth, with abdominal pain and dyspnea, had hemorrhaged severely from petechiae in and behind the abdominal cavity. Mrs. Chodo, who died on the twenty-ninth, had a hemorrhage in the wall of her heart. It was greatest where nerve impulses originated. Mr. Sakai, who also died on the twenty-ninth, had had severe shortness of breath. On autopsy he had a large hemorrhage in the chest and abdominal cavities, again in the presence of petechiae. Mr. Onomi bled to death from hemorrhages in his nose and rectum. Mr. Sakinishi, who died on the thirtieth in delirium, had massive hemorrhage in his chest cavity. Both lungs

were involved and petechiae were found in all the internal organs. Since his family insisted that we not remove his brain we could only surmise that brain hemorrhage had occurred.

Hemorrhage was the cause of death in all our cases. The extent and severity of petechiae, or surface manifestations of hemorrhage, bore no relationship to the extent of hemorrhage in the internal organs. Nor was the extent of internal hemorrhage the same in each organ. One organ might be badly involved and another spared. We could find no organ that had a greater tendency to develop hemorrhage than another, and the only organs consistently altered were the livers and spleens. In every case, they appeared smaller than normal, particularly the spleen.

Until now, we had interpreted the low white count as characteristic of the disease, but it became obvious that this was only one feature of a disease that involved platelets as well. Absence of platelets was responsible for hemorrhage and hemorrhage was the immediate cause of death . . .

(147–9)

The amount of medical detailing is truly enormous in such accounts, almost as though the magnitude of the disaster and the invisibility of the toxin can only be 'measured' in the form of organs, tissues and bones. The victim merely delivers her/himself up as the material story.

James Yamazaki, who arrived in Nagasaki four years after the bombing, writes in *Children of the Atomic Bomb: An American Physician's Memoir of Nagasaki, Hiroshima, and the Marshall Islands* (1995):

We felt it would be important to conduct autopsies on any of the surviving children who might die while our study was under way, but we realized that this would be most unwelcome to grieving parents. With the quiet guidance of the midwives, the parents came to understand the potential importance of autopsies to all of them.

(76)

Domon Ken narrates the last few weeks and death of 11-year-old Kenji – who was a 5-month foetus at the time of the Hiroshima bombing – from leukaemia. Ken begins by admitting his curiosity about radiation poisoning:

I wished to photograph at least one internal medicine patient who would provide proof of the insidious, persistent character of the "nail marks of the devil."

(Kyoko and Selden 160)

Ken notes the history of the autopsy process in Hiroshima, and the concealment of the results:

> I also knew that the corpse of a *hibakusha*, one exposed to the atomic bomb, was autopsied within several hours after death. Before, the corpse of every *hibakusha* was carried to the Atomic Bomb Casualty Commission at Hijiyama for autopsy by the American side at American initiative. Around the time the Atomic Bomb Hospital was established, it was decided that corpses that were with the Japanese side would be examined by the Japanese side at Japanese initiative. However, even then the autopsy results and a set of dissected organs seemed to be sent to the ABCC after all
>
> (165)[2]

He also provides a vivid sense-impression:

> The autopsy ended in a little over an hour. The metallic sound of the electric saw that was carried from the autopsy room to the waiting room was cruel. It made me vividly imagine the saw's teeth relentlessly severing the spinal cord.
>
> (166)

Ishii Ichiro in the same volume also documents doctors pleading with the family of the victims to permit autopsy because 'it will be good for a wide range of people' (Kyoko and Selden 198).

The transformation of the bomb victims into specimens rather than patients is also depicted in the Akiko case in James George's transnational nuclear novel, *Ocean Roads*. Showing Caleb her scar, she tells him how during her childhood, once every two years she had to go for medical tests but offered nothing else: 'they did everything but treat me. Treat us' (52).

The Japanese nuclear subject is both specimen and spectacle. S/he emerges in the medical examination narrative and in the autopsy narrative, of the individuals or their family members who permit and receive the news of the autopsy. When Ken, Michihiko and Ichiro visualize, recall or document the autopsies of the atomic bomb victims, they are not attending to the dead *per se*, but constructing the subjectivity of the readers, viewers and themselves who are drawn into the performance of the autopsy. They solve a puzzle, they document the cause of death, and they discover patterns – of radiation effects. 'They', in this case, are the viewers, physicians, pathologists and even photographers, who in the process of solving the puzzles around radiation's effects construct a nuclear subjectivity – and this is the 'potential importance', as Yamazaki puts it, of the autopsy.

The autopsy treats the human body as an *object*, a *thing*, before violating it – and this is sacrilegious in many non-Western cultures (Klaver 71). The important point about the insistence and documentation of the autopsies of atomic bomb victims is that the dead are, despite being objects, emanating signs. Elizabeth Klaver argues:

> The lesions or wounds in the cadaver are the signs of pathological events that the medical examiner *reads* rather than hears. . . .The lesions or wounds of the dead body . . . should be treated as natural signs, and thus receivable while *at the same time* independent of a sender. . . .The cadaver does indeed participate in the autopsy simply by providing (readable) perlocutionary statements.
>
> (81–2, emphasis in original)

In the natural order of things, the victims' bodies decompose. However,

> Once the body has died, it begins its process of decay and eventual disappearance . . . The only practical way to subvert this process is to record the body in some way, to memorialize it in language, diagram, description, and . . . in testimony.
>
> (Klaver 90)

The autopsy report or the testimony 'stands between the living and the dead' (91) as a mediating presence. But, as Klaver's innovative argument suggests,

> The (viewing) subject can therefore approach the dead body more comfortably, because the dead body is (or becomes) absent. The consequences of mediation, then, suggest that it is possible to overcome the trope of the speaking dead, and thereby to reduce the challenge the dead body makes to the subjectivity of the viewer . . .
>
> Autopsy not only initiates a renegotiation of subject/object positions, but also offers to mediate the dead body as a "being-as-subject" tolerable to the subject. Nevertheless, I want to complicate further this discussion by showing that the position of subject is exposed to the possibility of self-scrutiny coming from the autopsy itself. Autopsy is not only a way of mediating the dead body and its savage glare, which is subjectively tolerable, but is also a way of looking at one's own self, which is less tolerable – the subject as subject of death . . . the dissected cadaver seemed to signal a scandal of the body, the body fragmenting to its (originary) incoherent and inchoate state, a realization that reflected back on my sense of subjectivity.
>
> (92–3)

That is, the autopsy is not solely about the body of the dead victim, it is also about the viewer's/reader's 'sense of subjectivity'. This is precisely the sense one gets in reading Yamazaki, Hachiya and others: the autopsy of the atomic dead contributed to their own sense of nuclear subjectivity.

There is, one could argue, another dimension to this nuclear subjectivity that emerges from the autopsy narrative. The category of 'survivor' was itself a matter of some debate, as Susan Lindee notes:

> The number of survivors calculated also depends on the length of survival required for inclusion. "Survivors" could include those who were not immediately killed by the blast but who died hours or days or weeks later. Most estimates count as survivors those who were still alive three or four months after the bombings.
>
> (8)[3]

Then, the future lives of the survivors becomes the matter of some debate in the memoirs, based on the discoveries in the autopsies and medical examinations of the specimens, and the interpretation made of the extent of damage from radiation:

> The most serious clinical sign of radiation sickness is a decrease in the white blood cells, and pathologically, great changes were found in the hematopoietic system, especially in the bone marrow. Those who received fatal injuries have died within the past month. Patients with low white blood counts who survived this period are now stable or convalescing.
>
> (Hachiya 172)

Proceeding from these medical studies, Hachiya and others 'denied the rumors that Hiroshima would be uninhabitable for seventy-five years' (173). In the process, what they were constructing was a nuclear subjectivity around (i) the possible long-term effects of radiation, (ii) their fears of being victims of the radiation and (iii) a longevity defined by various medical problems and conditions. In other words, the medical studies of victims enabled Hachiya and others to generate a specific model of post-bombing lives for themselves. Hachiya informs us that his summary of findings were taken by a journalist and 'published almost verbatim' (180), indicating a larger readership for the earliest medical studies of radiation sickness – and, it could be conjectured, contributed to the fashioning of a nuclear subjectivity among the readers, hinging upon the *possibilities* of sickness and recovery.

For those with at least partial and *post-facto* medical knowledge at their disposal, their subject formation is almost entirely forged within the crucible of nuclear energy and atomic bomb testing. A case in point would be the 'downwinders' of Utah.

'Downwinders' were residents who lived 'downwind' from the sites of atomic testing, and deemed to be disposable people. In *Doom Towns* (2016), Andrew Kirk and Kristian Purcell draw the condition of life in

Utah after the Upshot-Knothole tests of 1953. The sheep are dying mysteriously, and the lab results show, as one officer admits, 'radiation was at least a contributing factor' (81). One such downwinder, Victoria Burgess, writes in her narrative of 2012:

> It means we lived downwind from the atomic bombs being tested by the government of our country. As a result of this exposure, our bodies were exposed to lethal radiation at a very young age.
>
> (Burgess, unpaginated)

Burgess recalls growing up watching the explosions and being thrilled by their 'power and beauty'. She then writes:

> at the reunion of my High School Class of '63, a high percentage of our classmates were suffering with serious illnesses or were dying and we wondered why. By our fortieth class reunion in 2003, one third of our classmates had died from radiation poisoning and various related diseases as a result of the atomic bomb testing.
>
> (unpaginated)

She cites the Atomic Energy Commission's report which 'justified their decision to release the atomic bomb near our area because we were "low functioning members of society"'. Years later, she is diagnosed with cancer:

> Radiation and chemo therapy was all the doctors of Western Medicine knew. I declined it because I believe dying of cancer from radiation is better than being treated with more radiation.
>
> (unpaginated)

Her entire life, as she ages, is built around battling the long-term effects of radiation to which she was exposed as a child:

> I feel helpless, hopeless, discouraged and depressed. My breast still hurts from the surgery, and nothing in Western Medicine can take the effects of radiation out of my body. Thus Cancer will likely kill me – no matter what I do.
>
> (unpaginated)

Then, Burgess makes common cause with other such victims:

> I find it interesting that the study compares us to the survivors of Hiroshima and Nagasaki Japan. Look what we did to those survivors and then we turned around and did it to ourselves. Coincidentally, Hiroshima and Nagasaki were bombed in 1945, the year of my birth!
>
> (unpaginated)

Sara Penny, another such 'downwinder' in Utah, begins her account thus: 'We knew we could die any day from about 5th grade', before listing all those she knew who died of cancer and fallout related illnesses in her childhood. Of her father she says:

> My father got an upper intestine cancer, which killed him and my mother had continuing health problems, including thyroid problems, which "may" have been caused by the fallout.
>
> (undated, unpaginated)

Pondering on her use of the word 'may', Penny writes:

> That "may" word is the upsetting word because you can't be definitely sure that all of this mayhem is from the fallout.
>
> (unpaginated)

Documenting the 'special needs' children born in the area, her husband's and her own ailments, and the US government's lies about testing, Penny writes:

> We live each day as though it may be our last because it really may be.
>
> (unpaginated)

Penny also makes a reference to Chernobyl – and therefore, like Burgess, making a common cause with other such nuclear subjects from around the planet.

The secrecy around the test and the non-sharing of information afflicted the US personnel involved in tests as well. For example, sailors on US navy ships and personnel islands in the Bikini area were also irradiated.

> Twenty-eight American service personnel stationed on Rongerik atoll to provide hourly weather reports were also exposed to radiation, and were not told when Bravo would be exploded. It was two days before the Navy arrived to pick up the Rongelap islanders and the US personnel – two days in which they breathed, slept and ate the fallout.
>
> (Dibblin, unpaginated)[4]

Rongelapese such as Jabwe Jojur were infuriated by the absence of information about the exact nature of the tests and what these tests produced:

> After three days we had burns all over our bodies, and our hair began to fall out: some people actually went bald. When we asked the Atomic Energy Commission doctors to help us understand what had happened, they did not tell us, and today they do not tell us the truth about our problems.
>
> (Dibblin, unpaginated)

Scientists like John Clark in their account of the Bikini tests admit that they did not know enough about possible fallout:

> Nor could we tell beforehand exactly how extensive the air-wave and tidal-wave effects would be or the precise amount and distribution of the "fallout"– the radioactive particles from the nuclear cloud which drop back to earth. In the business of testing nuclear devices there are always a few unknowns.
>
> (Clark, unpaginated)

Later, Clark writes:

> It was estimated that fallout radiation outside our blockhouse was several hundred roentgen. Had we been forced to stay outside the entire day without protective cover, it would have been fatal to all of us.
>
> The twenty-three Japanese fishermen in the Fortunate Dragon, which was seventy miles further away from the shot than we were, received burns. Twenty-eight American personnel manning weather stations, and 236 natives on Rongelap, Rongerik and Utirik also received radiation during the unforeseen fallout . . .
>
> But for the remainder of the tests on that atoll we made a change in plans--the firing operations were conducted from the command ship. Being guinea pigs once was more than enough for us.
>
> (Clark, unpaginated)

Clark and his compatriots chose to exit the islands before the next tests: he makes no mention of the islanders who remained trapped as the tests continued.

The Nuclear Community

In John Hersey's *Hiroshima*, he notes how 'wounded people supported maimed people; disfigured families leaned together' (unpaginated). As an illustration of the emergence of such a community, we turn to Takashi Nagai, the well-known doctor-survivor. Nagai in fact identifies three 'stages' in the lives of the Nagasaki residents in the aftermath of the bomb. In the first stage, lasting a month, people gathered and lived wherever they could ('the communitarian stage'). In the 'lost stage', people searched for their friends and families. From the second to the fourth month after the bombing, they began 'preparation for a new life' and took the 'first steps towards reconstruction' by pooling money and resources (112–3). Nagai consciously identifies the people of Nagasaki as a community having been 'chosen for the sacrifice', out of which 'peace was given to the world and freedom of religion to Japan' (in his 'Funeral Address for the Victims of the Atomic Bomb', 109), and thus underscores collective responsibility and sacrifice in a strange inversion of the victim image. Jessie Lennon speaks of how her entire community suffered from assorted ailments, principally cancer, after the Maralinga tests. As Lennon puts it: 'We still think about it. We still talk about it' (132).

The *hibakusha* writers, recognizing that they had been treated as less-than-human by the bomb, however, refused to be identified solely as victims, demonstrating, instead, resilience and resistance. Some in fact highlighted their animal-lives signalling, therefore, 'human "animality" . . . as signifying a political ontology of the weak' (Broinowski 97).

This political ontology of the weak relies on the development of an 'atomic gaze', as Broinowski calls it. Such a gaze is the departure point for an entire process of witnessing. Nuclear subjectivity is founded on intersubjectivity and concomitant *witnessing* of the horrors of war, injury and suffering, in the acknowledgement and apprehension of *shared* vulnerability, even as the protagonists in the narratives accept differences in the *levels* of vulnerability. This means, nuclear subjectivity as intersubjectivity allows the survivors and eyewitnesses to not only 'loosen the bonds of individual subjectivity within hegemonic normativity in order to succour the dead for the living' (Broinowski 104) but also develop a communitarian identity and subjectivity. I choose the term 'witnessing' over 'recognition' following the work of Kelly Oliver (2015) who argues:

> Witnessing takes us beyond recognition to the affective and imaginative dimensions of experience, which must be added to the politics of recognition. Perhaps this is why Butler talks about recognition in terms of "seeing as." Seeing as requires not only re-cognition but also imagination. Avowing the suffering of others caused by my own privilege, however, requires more than cognition or even imagination. It requires *pathos* beyond recognition.
>
> (475, emphasis in original)

This is the foundation for an ethics, argues Oliver, although there remains the 'tension between the concrete social, historical context of situated subjects, on the one hand, and the witnessing structure of subjectivity constituted through address and response, on the other' (475).

When Hara Tamiki, who survives the bomb, begins to make his way across the city, he is eyewitness to the destruction: bleeding and injured people, ruined buildings and fires.

> as far as the eye could see, buildings had collapsed, and only telephone poles still stood; the fire was already spreading.
>
> (Minear 49)

From being an eyewitness, Tamiki, in the space of a few lines in his 'Summer Flowers' (1949), moves to the role of *bearing witness*:

There was nothing left to fear; I myself had survived. Before, I had given myself an even chance of dying; now, the fact that I was alive took my breath away.

I thought to myself: I must set these things down in writing.

(49)

Oliver defines the difference between the two subject positions – eyewitness and bearing witness – in the context of the Jews in the camps:

As an *eyewitness*, she testifies (incorrectly) to the events of that particular day when prisoners blew up a chimney. In addition, however, she *bears witness* to something that in itself cannot be seen, the conditions of possibility of Jewish resistance and survival.

(Oliver 484, emphasis in original)

When Tamiki takes his decision to record what he had witnessed and experienced, he also asserts his (subjective) necessity to record other possibilities – of the resilience, community-work and support systems that the survivors almost at once put into place. Thus, Tamiki begins recording the various acts of rescue he and other fellow survivors engaged in, in the face of the fire, burning buildings and general ruin.

Thus, the act of bearing witness begins when the nuclear subject moves beyond the self to examine the other humans, and acknowledging that humanity as s/he knew it, had been eroded beyond recognition. Tamiki records:

Everything human had been obliterated – for example, the expressions on the faces of the corpses had been replaced by something model-like, automaton-like. The limbs had a sort of bewitching rhythm, as if rigor mortis had frozen them even as they thrashed about in agony.

(58)

Later, he reports what another survivor tells him:

On a ladder leaning on the riverbank, there were three corpses; rigor mortis had frozen them with their hands on the ladder. In a line waiting for the bus, corpses were standing just as they had been; they had died with their fingernails sticking into the shoulder of the person ahead of them in line. He also saw a large group of corpses-an entire unit of the labor corps mobilized from the countryside to clear firebreaks had been annihilated. Those scenes still did not equal the West Parade Ground. That was a mountain of dead soldiers.

(60)

The anonymity and the loss of identity is what one bears witness to. Echoing this theme when reading *Hiroshima Mon Amour*'s opening sequence, Yuko Shibata writes:

What is on display in this museum scene overlaps with this new perception of a human being left in the ruins after the atomic bomb attack. On exhibit in the museum, the faceless mannequins take on a twisted posture.

(77)

And yet, as Tamiki proposes, it is from/in this context of 'mountain[s] of dead soldiers' and automaton-like dead bodies and the dying, that he discovers a new humanity and its possibilities arise:

> No longer was there fear of air raid; now the broad sky wore an air of deep tranquility. I felt almost like a new person, someone born with that atomic thunderclap.
>
> (Minear 62)

It is not an easy subjectivity – it is one built upon a certain affective response to the loss and the suffering, and a resilience that comes from the need to survive and cope. As Tamiki puts it in the immediate sentences following the 'new person' passage:

> All the same, what of the people who died desperate deaths that day on the riverbed near Nigitsu and on the riverbank by the Izumi Villa?– I enjoy this tranquil view, but what has become of those charred ruins? . . .
>
> Even people about us who seemed fine at the time died thereafter of blood poisoning, and I was *haunted by a stubborn and incomprehensible unease*.
>
> (62–3, emphasis added)

The nuclear subject in Tamiki is one who is at 'unease', the result of having borne witness to the destruction of other humans (Tamiki mentions the death of the fish in the rivers too). He moves, in short, beyond his self, situated as he is within the context of the destruction of fellow-beings. His subjectivity, of being at 'unease', is determined by his historical location. As Oliver puts it:

> Subject positions, although mobile, are constituted in our social interactions and our positions within our culture and context. They are determined by history and circumstance . . . although subjectivity is logically prior to any possible subject position, they are always profoundly interconnected in our experience. This is why our experience of our own subjectivity is the result of the productive tension between finite subject position and infinite response-ability of witnessing.
>
> (Oliver 483)

Tamiki in fact tries to see the world around him through the eyes of the others:

> I saw rice paddies stretching all the way to the foot of the low moun-tain range. Tall green rice plants quivered under the hot sun. Was this rice the fruit of the land? Or was it there in order to make people hungry? Sky, mountains, green fields: in the eyes of hungry people they might as well not have been there.
>
> (63)

His sister Yasuko, likewise, is a nuclear subject entirely constituted by memories of the deaths of others (and not only through survivor guilt):

> As if still not wakened from the nightmare of that day, shaking like a leaf, she kept recalling that instant in great detail . . .
> Even now Yasuko trembled when she remembered so vividly a neighborhood child she had seen pinned under. It was a child in her own child's class who had taken part in the mass evacuation to the countryside; but the child had been simply unable to get used to life there, so it had been sent home to its parents.
>
> (Minear 63)

Tamiki, ensuring that neither he nor Yasuko present themselves as unique, writes immediately after this account:

> Many stories of this kind were making the rounds. When the bomb fell, Jun'ichi was pinned under but squirmed out, stood up, and recognized the face of the old woman of the house across the way, also pinned under. Though his first impulse was to rush to her aid, he could not turn a deaf ear to the voices of the schoolgirls screaming over at the factory.
>
> (63–4)

If, as Adam Broinowski argues, the political ontology of the weak and the atomic gaze involves 'empty[ing] the body of "self"' and instead to 'becom[ing] receptive to otherwise neglected, abandoned or silenced selves' (100), then this is precisely what Tamiki documents: a reaching out of the self towards something more.

Ōta Yōko in the Preface to the 1950 edition of her 'City of Corpses' informs us:

> I wrote *City of Corpses* between August 1945 and the end of November 1945. I was living at the time on a razor's edge between death and life, never knowing from moment to moment when death would drag me over to its side.
> After August 5, when the unconditional surrender of Japan ended the war, and after the 29th, alarming symptoms of atomic bomb sickness suddenly began to appear among those who had survived August 6, and people died one after the other.

I hurried to finish *City of Corpses*. If like the others I too was dying, then I had to hurry to finish it.

(Minear 157)

The writing is about bearing witness at a time when one's self and its continuity is itself under threat. The *need* to bear witness is concomitant with a sense of imminent ending – and this is the nuclear subjectivity of the novelist/writer. Later in the Preface, Yōko admits to the inadequacy of her work, her vision and even her style:

As I read, I could not help feeling, all the more keenly, that what I experienced was small and insignificant when set against the extraordinary suffering that unfolded throughout Hiroshima on August 6, 1945.

My pen did not take in the whole city. I wrote only of my very limited experience . . .

The whole city was buried in a calamity more sad and severe than the scenes I saw on the riverbed and in the streets: that fact I should like my readers to be aware of.

Readers will probably find my style unsatisfying. In rereading this book today, five years later, I myself felt impatient at many points. There arose before my mind's eye the conditions in Hiroshima then, conditions I was unable to describe adequately; I could not help remembering physical and spiritual suffering so severe it seared my very soul.

(148)

She goes on to speak of the difficulty in describing what she witnessed (148) and her determination:

But precisely for this reason it was all the more important that I write about Hiroshima. The disaster of Hiroshima cannot be considered apart from its historical significance. When one realizes this, then even in a work of literature one cannot fabricate, one cannot take one's time. One should transplant the situation into fiction, preserving its factual underpinnings.

(149)

The nuclear subject perceives him/herself in terms of the other humans with whom s/he had been connected – a connection the bomb severed. In Tōge Sankichi's poem, 'Prelude', he exhorts:

Bring back the fathers! Bring back the mothers! Bring back the old people!
Bring back the children!

> Bring me back!
> Bring back the human beings I had contact with!
>
> (Minear 305)

In 'August 6', Sankichi speaks of the 'soul-rending appeal' in the eyes of the dead lying around: it is an appeal to the response-ability of the nuclear subject. In 'Eyes', likewise, Sankichi's speaker *feels* appealing eyes on him:

> Eyes fastened to my back, fixed on my shoulder, my arm.
> Why do they look at me like this?

The eyes look at him because he is intact:

> Erect, clothed, brow intact and nose undamaged,
> I walk on – a human being:

He believes they groan his name, as he seeks his family in the heap of injured and dead (317–8).

The speaking subject in Sankichi – the 'I' – is constructed only within the act of listening-responding to the injured and the dead, even when the voices calling out to him or eyes staring at him are entirely imaginary. To phrase it differently, the subject is the effect of a relational process – of address by the atomic trauma victim and response by the survivor-speaker. The speaker, whose own self has been subject to the trauma of the bomb, continues not through a process of interiority or self-recognition but through the affective response-ability to the Other, 'between finite subject position and infinite response-ability of witnessing', as Oliver, quoted above, puts it.

Reconstructing the Human

Kenzaburo Oe in his Prologue to *Hiroshima Notes* cites a letter he received from a Hiroshima resident on reading his (Oe's) essays in the newspaper. The letter-writer said:

> People in Hiroshima prefer to remain silent until they face death. They want to have their own life and death. They do not wish to display their misery for use as 'data' in the movement against atomic bombs or in other political struggles.
>
> (19)

The letter-writer endorses the silence of the victims and agrees that they 'leave their testimony for the historical record' (20). Oe acknowledges the points made by the annoyed letter-writer and, while stating that Hiroshima cannot ever be expunged from the mind, reserves a note of appreciation for the victims: 'the perseverance with which the Hiroshima people bear their miserable solitude is not a dogmatic stoicism' (24). Later, in another essay, Oe speaks of Hiroshima people such as Dr. Shigeto:

> The doctor is himself an A-bomb victim; he, too, witnessed that hell. He is a typical Hiroshima man who keeps up the fight against the A-bomb effects that even now remain deep in human bodies.
>
> (48)

Through these people, he writes, he met 'the true Hiroshima' (56). Oe records how, as children die of inherited diseases – the effect of radiation poisoning – as do the adults, a new humanity seems to rise. Oe speaks of a girl in her teens who 'fell in love, married, and had a baby'. Oe comments: 'I think such courage in the face of desperate anxiety may be called truly human' (58). He praises their patience (74) and their efforts at medical reviews and examinations to document the effects of the atomic bomb (77).

The subjects here refuse to participate in their continuing victimhood and datafication, as Oe notes admiringly. The victim-survivor does not engage with the larger demands of the reparation-apology-memorializing process, as the Hiroshima survivors Oe speaks of, do.[5] In the process, Oe also flags the politics of memorialization which demands the participation of the 'miserable victim', so to speak, which would also then employ the *hibakusha* as a symbol or place-holder for the values, expectations and hopes of anti-nuclear protestors. The Hiroshima peoples' silence prevents them from being displaced onto a 'Hiroshima avatar' or type which represents *other* people's aspirations and politics. Oe is also indicating that such a displacement would have altered their subject-nature, their subjectivity, and instead objectify them once again (they had already been objectified when the Americans decided that their lives as humans did not matter enough).

Now, the nuclear subject in contemporary Japan *may* be an instance of the 'deterritorialized modern subjectivity' wherein the 'average daily life in the context of globality is shaped by structures, processes and products that originate elsewhere' (Heise 54) – in this case, nuclear power and the bomb produced and employed by the USA. But what Oe does is to reclaim the subjectivity for some of the Hiroshima *hibakushas* by demonstrating how they do not fit into the global model and instead reterritorialize themselves through their localization of memories and their dignity of silence.[6]

Oe's work, produced in the decades after the bombings, is focused on a nuclear subjectivity founded on resilience and reconstruction of the Japanese, which he believes constitutes the foundation of a 'new humanism'. In his chapter 'The Moralists of Hiroshima', Oe cites numerous oral histories and interviews of the courageous hibakushas before elaborating on the humanism that emerges from their views, practices and lives.

> If a mother wants to see her dead deformed child so as to regain her own courage, she is attempting to live at the minimum limit under which a human can remain human. This may be interpreted as a valiant expression of humanism beyond popular humanism – a new humanism sprouting from the misery of Hiroshima.
>
> (83)

This courage, which marks a 'people who go on struggling toward a miserable death' (Oe 95), is defined by him as 'the thoroughly and fundamentally human sense of morality in the Hiroshima people 'who do not kill themselves in spite of their misery' (84). He is all praise for the survivors who, despite the 'look in the[ir] eyes . . . not so much grief or anger as it is . . . a kind of shame' (91), go on living and do not pity themselves. Oe also sees this as a 'struggle to gain new life . . . to struggle along the way to miserable death, or until meeting a miserable death' (95). This is at once survivor subjectivity and nuclear subjectivity that forwards an entire new form of the human.

When Oe concludes this particular essay, he wishes for a universal model of such a subjectivity in the event of another nuclear holocaust. He quotes Hara Tamaki's 'Give Me Water' in which Tamaki speaks first of 'the moaning of a man' and concludes with

> Scorched, smarting;
> The ruined face of man!
> (Oe 129)

Here, the poet's move from 'a man' to 'man' where the first instances a singular man's moaning but the second instances a universal 'smarting' and scorching is significant because it gestures at the possibility of universal annihilation – a lesson learnt by recognizing the power of the atomic bombing of Japan.

Moving beyond the immediacy of Japan and 1945, Oe writes:

> If ever we experience another massive nuclear flash and thunder over our heads again, I am sure that the morality for survival when surrounded by death and desolation will need to draw on the wisdom of those who, through their bitter experiences in Hiroshima, became the first moralists, or 'interpreters of human nature', in our nuclear age.
>
> (96)

He concludes the essay thus:

> No doubt the preparation of a white paper on A-bomb victims and damages, to mark the twentieth anniversary of the atomic bombings,

would contribute much to the preservation and promotion of that precious morality.

(96)

This 'precious morality' would be/ought to be the hallmark of nuclear subjectivity for the world at large, suggests Oe. Oe contrasts this humanism, born of unmitigated, long-term suffering with the 'Western' model of humanism:

> I have a kind of nightmare about trusting in human strength, or humanism . . . My nightmare stems from a suspicion that a certain 'trust in human strength', or 'humanism' flashed across the minds of the American intellectuals who decided upon the project that concluded with the dropping of the atomic bomb . . . I suspect that the A-bomb planners thought in such a way; that in making the final decision, they trusted too much in the enemy's own human strength to cope with the hell that would follow the dropping of the atomic bomb. If so, theirs was a most paradoxical humanism.
>
> (115–16)

Having contrasted the two humanisms thus, Oe turns to the 'human strength' of the Japanese the foundation of their (re)new(ed) subject position as A-bomb survivors.

'Human goodwill', writes Oe, 'began to go into action as people made their first moves toward recovery and restoration', 'even while the smoke still rose from the wasteland of total destruction' (114–15). He then turns Hiroshima peoples' survivor subjectivity into something truly unique. Their efforts to 'recover and rebuild' was of course for their own sake but 'doing so served also to lessen the burden on the consciences of those who had dropped the atomic bomb' (117). Oe writes: 'it is the attacking wolf's confidence in the scapegoat's ability to set things straight after the pitiless damage is done. This is the nightmare I have about humanism' (117). He gives an example:

> The patience of the A-bomb victims quietly awaiting their turns in the waiting room of the Atomic Bomb Casualty Commission . . . At least it is true that their stoicism greatly reduces the emotional burden of the American doctors working there.
>
> (118)

Oe is indexing not the humanism of the nuclear subject alone, but the morality and politics of resilience which is the core of the subjectivity.

Apocalyptic Knowledge

The reconstruction of the human and human subjectivity after the bomb also produces and is produced by a certain amount of knowledge and clarity about the world, technological, medical, moral and political, as select survivor texts demonstrate. 'Apocalyptic knowledge', writes Russell Meeuf, 'is only produced through violent spectacle' (286). The nuclear subject as it reconstructs the human of/for the future is imbued with a nuclear apocalypticism:

> Nuclear apocalypticism . . . is more complex as a kind of knowledge system than its apocalyptic forbearers because of its insistence on a cataclysmic end not only to humanity but to the possibilities of knowledge formation itself.
>
> (Meeuf 287)

One can infer a nuclear subjectivity founded on such apocalyptic knowledge in Oe's account of physicians and medical staff in the decades following the bombing, but also other survivor accounts. He has already offered a clear vision of the comparative humanisms of America and Japan, as discussed above, a vision that emerges from an understanding of the politics of the bombing (he uses as a hypothetical test case, an atomic bombing of the Congo, 116).

The Hiroshima City Medical Association in 1958 distributed a questionnaire to the surviving A-bomb victims among its doctors, and later printed it in the Hiroshima A-bomb Medical Care History. The questionnaire included questions on the nature and extent of the doctors' own injuries as well. Oe observes:

> If the awesome impact of the bombing had robbed any Hiroshima doctor of all desire to engage in rescue work, it would not have been, humanly speaking, particularly abnormal. But after receiving the questionnaire, such a doctor could hardly have slept soundly.
>
> (120)

Now, the knowledge that 'not only were people suffering now that the war had ended but also that they would continue to suffer for many years to come' (124). One young dentist, who was a part of the rescue work although he had fractured both his hands and extensive burns on his body, hanged himself after this knowledge came to him. Oe is signalling the price of apocalyptic knowledge here. Other doctors, notes Oe, recognised radiation sickness in the form of, for example, leukaemia. But, the doctors 'simply refused to surrender' (131).

After Hachiya returns to his practice – he is injured in the bombing – he observes the patients and their deterioration, often documenting his doubts like 'why should leukopenia occur?' (98). He records the recognition of diseases of the blood occurring in hundreds of patients coming in, 'unusual symptoms' like losing hair (102), among others. They discover over time a link between 'distance [from the hypocentre of the bomb blast] and white blood count' (109). Through such tedious work, the atom bomb, which he calls 'this unknown enemy' (109) then becomes more decipherable to him and his fellow-doctors. 'People who now appeared well should be on their guard', he concludes (112). The inevitability of many deaths from radiation sickness is a truth that Hachiya and his colleagues come to understand in the months following the bombings, and this truth is a part of their reformed nuclear subjectivity as well, as professionals faced with an impossible task.

Takashi Nagai writes that he and his colleagues in Nagasaki were 'dedicated to the truth [about the bomb and radiation sickness]' (73). He writes:

> For the first time in history the atom had exploded over the head of human beings. Whatever symptoms might appear, the fact was that the patients we were now treating had diseases that were completely new in the annals of medical history. To ignore these patients would not only be an act of cruelty toward individual persons, it would be an unforgiveable crime against science . . . we ourselves were already experiencing in our bodies the first stirrings of atomic sickness. If we continued our rounds without adequate rest, our symptoms would get worse and worse . . . even if we didn't die, we would certainly fall seriously ill.
> And yet my academic conscience gave strength to my body.
>
> (73)

Nagai's apocalyptic knowledge-making is at once grounded in his own body but also in the institutions that make him a professional: the medical college and the university. He has already informed us that the institution – the Nagasaki School of medicine – 'was reduced to ashes' (43). In the face of this apocalyptic scenario, the principles of medical knowledge-seeking remain well embedded in the practices that Nagai and his colleagues continue in the face of terrible odds. Although such apocalyptic knowledge comes from violent spectacle as noted above, it is directed at a future but not necessarily towards the end of the world.

In Chernobyl, the workers who shovelled earth trying to bury the contamination had acquired an understanding – albeit too late to save themselves (because 'the truth about the high doses they were receiving was concealed from them', Alexievich) – of what had happened. Svetlana Alexievich writes:

> I realized that they were consciously converting their suffering into new knowledge, donating it to us. Telling us: mind you do something with this knowledge, put it to some use.
>
> (30)

Alexievich is gesturing at apocalyptic knowledge – the workers first on the scene, and many others later, would die from radiation sickness in months – derived from personal experience, but which implicitly exhorts the survivors of Chernobyl to make something of the knowledge. It is, in Alexievich's view, an exhortation to transform our views on nuclear power itself. The nuclear subject, imbued with the knowledge gleaned from the sufferings of others who had died, ought to feel a compelling sense of obligation to use the knowledge for the betterment of mankind. She cites Valentin Alexeyevich Borisevich, former laboratory director of the Institute of Atomic Energy, Belarus Academy of Science:

> Why is there so much interest nowadays in an alternative reality? In new knowledge? Man is breaking away from the ground . . . He is operating with different categories of time, and not just with the earth, but with different worlds. Apocalypse. The Nuclear Winter. The explosion of a large number of nuclear weapons will cause an enormous conflagration.
>
> (Alexievich 221)

Borisevich – who has cancer – concludes that only the time of 'our lives' has any meaning (Alexievich). In each of these instances, the nuclear subject is constituted through apocalyptic knowledge, but this is also knowledge s/he wants to disseminate so as to construct a nuclear subject *who is not limited to geographical locations*. For example, Lyudmila Dmitrievna Polyanskaya links multiple sites:

> I realized that Chernobyl was more remote than Kolyma, Auschwitz and the Holocaust. Am I making sense? Someone with an axe, or a bow and arrow, or a man with a grenade launcher or gas chamber, could not kill everyone, but a man with the atom . . . That means the whole world is in danger.
>
> (Alexievich)

Polyanskaya is a nuclear subject whose apocalyptic knowledge originates in an understanding of the global history of man-made destruction, in contrast with certain kinds of nuclear subject who remains localized, as we have seen in the case of Oe's quotations from and reports on the Hiroshima *hibakushas*.

Uncanny Nuclear Subjects

In one of the most cited poems of the nuclear age, Hone Tuwhare's 'No Ordinary Sun', the speaker addresses a tree. The tree, says the speaker, need not raise its hands in supplication, because the sap inside it will not rise again with the morning sun. The tree can no longer serve the lovers with its shade, or provide the birds with shelter. It then goes on to describe the reasons why: the flash is no 'gallant monsoon's flash', nor is it the force of the trade wind. Neither is this brightness emanating from the sun, because 'this is no ordinary sun' (23).

The poem has come in for considerable attention. Elizabeth deLoughrey argues that 'the poet relies on the metaphorization of the bomb to establish an allegory about the sun and tree' (248). Further, the poet naturalizes the flash of the bomb's blast and denaturalizes the natural winds, tides and brightness (249) even as the poem ends in what she calls 'total light', ' but illumination does not follow – this world is "shadowless" and "drab"' (250). Jessica Hurley suggests that the poem captures 'an experience of the forced defamiliarization of the everyday world by the colonial and neo-colonial nuclear infrastructures that produce unevenly distributed harm in the present' (95), and 'Tuwhare represents the nuclear uncanny mimetically, by representing the defamiliarizing of the everyday world through nuclear violence' (98). Michelle Keown argues that 'the detonating bomb appears as a "monstrous sun" that, rather than fostering growth through its light and warmth, creates apocalyptic destruction' (591). Hurley's reading, which resonates with my own, points to the nuclear uncanny as a feature of work that shows how the familiar world has been rendered a little strange due to the bomb.

Joseph Masco (2006) was one of the first commentators to discuss the defamiliarization of the everyday in places affected by nuclear testing. Masco writes how the natives began 'inhabiting an environmental space threatened by military-industrial radiation' (28). The 'psychosocial effect of nuclear materials' was to

> render everyday life strange, to shift how individuals experience a tactile relationship to their immediate environment. This gets at the root definition of the uncanny as *Unheimliche* [sic], or the unhomely, for the invisibility of radiation can make any space seem otherworldly, strange, and even dangerous. Indeed, what could be more 'unhomely' than the introduction of nuclear materials into one's everyday environment or body?
>
> (33–4)

Their homes and their ecosystems transformed irrevocably, writings by those who were in the zone of nuclear testing and explosions experience the nuclear uncanny. Thus, in Tōge Sankichi's 'At the Makeshift Aid Station', the speaker, looking at the severely burned girls (who no longer have eyes or lips to cry with), wonders if they understand 'how far transformed from the human they are' (Bradley 15–6). The uncanny is not simply experienced in this transformation of the body into something strange, but also in the impossibility of their family members identifying any of the injured:

You are simply thinking,
thinking
of those who until this morning
    were your fathers, mothers, brothers, sisters
(would any of them know you now?)

                                     (Bradley 16)

The poem concludes with the speaker assuming that the dying girls lie there 'thinking/of when you were girls, human beings' (16). The foreign, then, is not out there, but closer home, in the very ontology of the nuclear victim and the familial relations – which were taken for granted – that were constitutive of the identity and subjectivity of the person before the bomb. (This absence of recognition by the family would be repeated in Sankichi's 'The Shadow', Bradley 22–3.)

Their everyday space consists of something foreign *in* their selves and in their settings (the 'revelation of something unhomely at heart of hearth and home', Royle 1). But beyond this defamiliarization of the everyday – as seen in Tuwhare's rendering of the sun and the wind, or nature itself, into something else altogether – which the above commentators point presciently to, the nuclear uncanny possesses other features worthy of attention.

The uncanny is linked to an experience of *time*:

> The uncanny is tightly bound to temporality; the inability to return to past sites and past selves often comes into conflict with our memories of these pasts. Memories can become ghosts that haunt the present. The uncanny can be understood as the cohabitation of tenses, memories of a familiar past rubbing up against the strange newness of the present. Familiarity depends on the interaction of experience and recollection, a concurrence between one's perception of what was and what is. I experience the uncanny when my expectations, inevitably based on memory, are upset; when the familiar, the recognizable, is infiltrated by the strange, the unrecognizable, that is, when the past and present fail to align properly.
>
>   (DeFalco 9)

Take, for example, Jane Dibblin's documentation of the Marshall Islanders' experience:

> Our hair fell out. We'd look at each other and laugh – you're bald, you look like an old man.
>
>   (unpaginated)

The collapse (acceleration?) of the natural aging process due to the intervention of the nuclear fallout hastens the future into the lives of the people, and they are each rendered unfamiliar to the other as a result.

The defamiliarized terrain and nature, the everyday and the quotidian, are characterized by the past that spills into the present and looks forward to the future so that the event of the bomb or the time after the bomb/test is imbued with more than just a sense of the present. To return briefly to Tuwhare's poem, the speaker perceiving the tree, treats the tree's past – with the physiological processes of sap rising, or the instrumental use of the tree's shade by humans – with the sense of a future in which these processes and functions *will* end:

In the shadowless mountains
the white plains and
the drab sea floor
your end at last is written.
(23)

If Tuwhare *sees* the future in the nuclear present through the defamiliarized everyday, in Benjamin Alire Sáenz's poem on the Trinity test, 'Creation', he uses the myths, legends and stories of ancient civilizations, but also the everyday, to speak of mankind's newest 'creation'. The 'man-made flash' is 'twice as large as the sun' and the brightness of the moment is captured in the image of photography: 'photographed the moment/in fire' (Bradley 3). Given the absolute strangeness of the bomb, Sáenz can only resort to older familiar tropes to describe the singularity: the dust tower is bigger and taller than any tower the Aztecs built, the smoke was a stronger, stranger 'new incense' in which old modes of worship and ritual 'perished'. The hot air 'invok[ed] Indian winds'. Eventually, writes Sáenz, the sun 'no longer gave enough light' and those who witnessed the brightest flash ever on earth had 'grown accustomed to the dark' (Bradley 4). In the process of linking the past with the present and the future, Sáenz also provides another instantiation of the uncanny. Central to the uncanny is what Freud identifies as a return to unformed primitivism (Freud 393–7), and what Hélène Cixous refers to as its 'mythic anthropology', a 'foundation of gods and demons' (539). The uncanny is also connected, Cixous adds, to a series of anecdotal examples, literary and biographical mini stories (539). The return to ancient civilizations – a forgotten past – even as he draws attention to the contemporary horror that is nuclear power, enables Sáenz to speak of the defamiliarized present through the trope of something more recognizable, but which too is foreign because of its separation in time (Aztecs, the Native Americans and other cultures).

In William Stafford's well-known 'At the Bomb Testing Site', the opening image is of a lizard that 'waited for history' (Bradley 73). The 'history' is at once the history of the land and forthcoming history – the latter being scripted at Trinity on 16 July 1945. The lizard-as-witness is also unique because it had its eyes fixed on the future, 'farther off/than people could see'. The world, and the desert, are 'ready for change', writes Stafford. Stafford leaves it open-ended as to what the change may be: would it be annihilation? The lizard grips the desert hard, says Stafford, perhaps in suspense, bracing for whatever is coming: would it survive the change? The canniness of the lizard is in fact a preternatural knowing: it can see beyond the humans (life after humans?), and see beyond what the humans see. This canniness is matched by the possible strangeness of the coming change indicated in the vagueness of the phrase, '*something* might happen'. The change may be an 'important scene'. Stafford's invocation of the stage/film with the use of 'scene' also invokes the transient nature of the stage: the show must end at some point.

It is also, understandably, linked to the experience of space. After the bombing, Michihiko Hachiya writes:

> Many of the inhabitants I had known, but the place was so strange to me now, that for the life of me, I could not have said where any of them had lived.
>
> (50)

Visiting 'the Zone', as the contaminated area around Chernobyl was called, Gennady Grushevoy, member of the Belarusian Parliament, and Chairman of the Children of Chernobyl Foundation, observes:

> Everything was turned on its head, topsy-turvy. I realize that now. There was a strange sensation of death . . .
>
> (Alexievich 146)

The place and its inhabitants are familiar and yet rendered strange, known and no longer *knowable*. Dylan Trigg argues the uncanny's spatiality: 'places can, for instance, become singular in the library of our memories through their very unfamiliarity. Indeed, precisely through their strangeness, places become memorable by disturbing patterns of regularity and habit' (9).

Trigg also argues that the routine and the familiar creates the sense of the everyday space as unshakable, that our memory of the home and its material objects lends it the sense of stability and continuity:

> We can thus speak of memory in terms of its being as much bound with subjectivity as it is the materiality of objects in the world. In a word, the places in which we live, live in us. More precisely, those places live in our bodies, instilling an eerie sense of our own embodied selves as being the sites of a spatial history that is visible and invisible, present and absent.
>
> (33)

And further:

Things that we cherished as assuming a particular appearance – warm, imposing, intricate – tend to materialize as malformed, unsettled, overrun, and, in a word: alien. The world to which we had previously accustomed ourselves through memories and dreams now adopts a sinister presence, forever sliding in and out of our temporal frame.

(37)

The uncanny occurs when this stable frame quivers. When Mary Jo Salter describes the lives of people after the reactor accident in her poem 'Chernobyl', she focuses on the everyday spaces and practices that have been rendered bizarre. First, the milk – believed to be contaminated – has to be thrown down the drain. Then, the lambs are painted blue, 'not to be eaten', until the next spring when 'they'd grown into blue mutton' and, presumably, ready to be eaten. What is frightening is not these changes that render the everyday uncanny, although they are significant too. Salter concludes with the illusion of safety under which the Chernobyl populace lives:

> Safe
> and innocent, the rain
> fell all night as we slept,
> and the story at last was dead –
> all traces of it swept
> under the earth's green bed.
> (Bradley 100–1)

Salter spatializes the long-term effects of the toxic rain by pinpointing the people who sleep under it. Then, she employs the traditional role of rain by referencing the 'green' earth – where rain revives the earth, and encourages growth and life. However, the poem's invocation of this traditional trope – recalling Tuwhare's inverted heliotrope – is tragic-ironic as the lines imply. First, the 'story' is dead: a reference to the first stanza of the poem where 'headline of a story' is employed to suggest how Chernobyl made headlines, but also about the myths of safety and radiation levels that were circulated alongside cautionary tales about contamination. Second, the story is itself dead because no one bothers about contamination anymore, just as the painted lamb became the painted mutton for consumption. Salter's people 'have had enough' and have decided that they can no longer live in fear and so the stories are buried in the earth. With this image, Salter gestures at two things: the stories have been buried in the ground, implying the burial of nuclear waste in subterranean spaces (it was actually called the 'crypt' in several places, such as Bikini and Nevada). Then, the 'green' earth is misleading because the earth is contaminated. In short, the nuclear is grounded, spatialized, as stories and radioactive waste are both buried in the soil and what may emerge from this soil – rejuvenated by the rain – is a matter of horrific conjecture. The nuclear uncanny here is the experience of a routine agricultural process, but whose end results are likely to be a strange and toxic harvest.

Or take the narration of Marshall Island children who play in the radioactive sand thrown up by the Bravo test (1 March 1954):

> By mid-day the fallout, a fine powder which fell from the sky, had reached Rongelap. The children had seen photos of snow, and at first the young ones played in it.
>
> (Dibblin, unpaginated)

This is uncanny inversion of the space of everyday play, and snow and radioactive dust become mirror images of each other in the children's perception.

In Hiromi Kawakami's short story, 'God Bless You, 2011' (a reworking, after Fukushima, of her earlier 1993 tale, 'God Bless You'), workers in the paddy fields 'were encased in protective suits and masks with waders that extended to their waists' (unpaginated). The protagonist, who has just gone out on a walk with a person/bear (the tale does not make it clear whether this 'bear' is a person because she says: 'When he was talking, his voice sounded entirely human, but when he hemmed and hawed like this, or when he laughed, he sounded like a real bear') but without their protective suits so, 'a surprising number of cars passed us. They slowed to a crawl as they approached and made a wide circle around us' (unpaginated). The protagonist comments: ' "Maybe they're keeping a distance because we're not wearing protective suits," I said' (unpaginated). The fish in the river, the bear says, eat the moss at the bottom of the river, where a lot of Caesium also accumulates. When the story ends, the protagonist, now back home, writes in her diary:

> I recorded my estimate of the radiation I had received that day: thirty micro-sieverts on the surface of my body, and nineteen micro-sieverts of internally received radiation. For the year to date, 2900 micro-sieverts of external radiation, and 1780 micro-sieverts of internal radiation.
>
> (unpaginated)

This is an instantiation of the nuclear uncanny where the knowledge of radiation and its effects makes its absorption by the body a part and parcel of *everyday* record-keeping, transforming the body, its setting and its quotidian processes into something strange and unfamiliar.

Another way to think of the nuclear uncanny is to think of radiation following from nuclear testing – and the testing itself – as revealing some secret *within* oneself, or within one's irradiated body, to be specific. Nicholas Royle has argued that there exists a sense in the uncanny of a 'secret encounter', an 'apprehension . . . of something that should have remained secret and hidden has come to light' (Royle 2). It has to do, writes Royle, with 'feelings of uncertainty', especially about the 'reality of who one is and what is being experienced' (1). This sense of the uncanny can be detected in the theme of irradiated bodies. I have already cited Shinkichi Takahashi's 'Explosion' in which he speaks of his shrinking body, its implosion. Later in the poem, he would state the reasons for this shrinkage: the 'pollution' inside the marrow. Takahashi concludes the account of implosion with this 'scant blood left/reduced to emptiness by nuclear/fission' (Bradley 298).

The irradiated body's uncanniness stems from its metamorphosis into something radiant, even spectral. In the immediate aftermath of the bombing, the survivors caught by the blast are described as ghostly, abhuman in some texts. Here is Hachiya:

> There were the shadowy forms of people, some of whom looked like walking ghosts. Others moved as though in pain, like scarecrows, their arms held out from their bodies with forearms and hands dangling.

Instances of people 'dangl[ing] both hands in front of their chests ghostlike' abound in Kyoko and Selden (9, 16). People-as-ghosts are common tropes in the poetry as well. Nakamura On describes the 'procession of ghosts' in 'City in Flames' (Kyoko and Selden 118). Sakamoto Hatsumi, a schoolchild, in 'The Atomic Bomb' writes that when the bomb fell, the day turned into night, and 'people turn into ghosts' (Kyoko and Selden 127). Svetlana Alexievich records stories of people whose bodies became other-than-human and therefore foreign and uncanny:

> my son went to school. He burst into the house after his first day, crying. He had been sat next to a girl, and she complained she didn't want to sit there because he was 'radiated' and if she sat next to him she might die. My son was in fourth grade and, unluckily, he was the only person from Chernobyl in the class. They were all afraid of him. They called him the 'Glow-worm', or the 'Chernobyl Hedgehog'.
>
> (111)

Ghostification, a component of the uncanny, is the transformation of the human body into something else, familiar and yet strange.

For Masuji Ibuse's Shizuma in *Black Rain*, the bomb-exposed body has itself become an uncanny sight:

> Could this be my own face, I wondered. My heart pounded at the idea, and the face in the mirror grew more and more unfamiliar.
>
> (143)

Ibuse turns a pathological condition induced by the bomb into something uncanny and eerie when he writes of a victim, Iwatake, who reports a ringing in his ears:

> Today, I have one earlobe missing, and when I take a drink the scars on my cheek and wrists turn red, but apart from a stubborn ringing in my ear I have no after-effects at all. The one thing that troubles me is the ringing; it persists in my ear day and night, like the tolling of a distant temple bell, warning man of the folly of the bomb . . .
>
> (270)

A different uncanny also presents itself in nuclear literature. Mutated foetuses and births – the source of much horror in popular depictions of nuclear apocalypse – are recorded by people in the vicinity of atomic testing sites, such as Bikini Atoll:

> I saw three different women give birth to strange things after the 'bomb'. One was like the bark of a coconut tree. One was like a watery mass that was not humanlike. Another was again like a watery mass of grapes or something like that. I believe that these things are all caused by 'the bomb'.
>
> (Dibblin, unpaginated)

If the mutant foetus defamiliarizes the human, it suggests that the uncanny asserts itself at the opposite end of the life spectrum too. Thus, even after death, the irradiated body is a source of uncanny effects. For example, in Chernobyl, the residents of the Zone discover something frightening and spectral about the buried dead:

> there were various rumours around: even after death, Chernobyl victims were said to glow . . . At night, a light would appear above their graves.
>
> (Alexievich 291)

In John Wyndham's novel, *Re-birth,* there are mutant babies born in families of survivors of an unspecified catastrophe (it appears to have been a nuclear war). In texts like Wyndham's, the boundary between human and non-human other breaks down through the mutated bodies that are uncannily, eerily, human and yet not quite.

The theme of mutated babies and dead foetuses in atomic bomb writing reasserts the uncanny's temporal aspect already mentioned (the 'cohabitation of the tenses'): they index, materially, a failing human future. The traditional Gothic's fear of inheritance (Baldick 1992: xix) is replaced by the fear of the future where babies and children, both familiarly and yet not, represent the fate of the human race exposed to the bomb. The references to strange births and mutant forms reference the border between the 'normal' and the 'pathological', a border that has been called into question by radiation.

The nuclear subject, this chapter has shown, is constituted of multiple strands. The subject embodies an anxiety, an injury and a potentially uncertain future in a toxified landscape, from Nevada to the Pacific. Accounts of the nuclear subject thus encode an insecurity discourse of the future. The experience and spectacle of Hiroshima, Nagasaki and Chernobyl writ the power of this energy upon the human form, altering in many cases the intergenerational ontology of individuals and entire communities, as the toxic somatographies from these sites demonstrate.

The nuclear subject – and we are all, in a sense, nuclear subjects – is to be found in all parts of the world, as this chapter's examples have shown, from Nevada to the Marshall Islands, from Chernobyl to Fukushima.

Notes

1　Writing about *hibakusha* and memory in her book on American survivors of Hiroshima and Nagasaki, Naoko Wake notes that more women were 'available and willing to remember in public than men' and so 'breaking the silence around the bomb continued to depend on women's participation, even predominance' (223).

2　There were mixed responses to the Atomic Bomb Casualty Commission's activities in Hiroshima and Nagasaki. Many Japanese saw the ABCC as yet another invasion and bombing of their nation (see Lindee 1994: 13. For Japanese responses to the ABCC and its secrecy around autopsy materials, see Lindee 18, n. 3). The secrecy around the autopsy reports did not help matters either.

3　In her discussion of the ABCC, Lindee writes:

> The problems to be explored at Hiroshima and Nagasaki involved a recognized but poorly understood hazard to human health with long-term effects that were not immediately apparent in the survivors. These invisible effects could be revealed only through epidemiological research and statistical analysis.
>
> Radiation released by the bombs killed the residents of Hiroshima and Nagasaki in ways that initially baffled Japanese physicians. Many survivors who seemed to come through the blast unhurt began, within minutes, hours, or days, to manifest the symptoms of acute radiation sickness. Sudden, severe nausea and diarrhea were the first signs, followed later by subcutaneous bleeding and gingivitis (inflammation of the gums). Many of these seemingly unhurt survivors died in the days and weeks after the bombings . . .
>
> (11)

4　In a savage irony, the *Lucky Dragon*, caught in the Bikini Atoll test (1 March 1954) fallout, was treated as a spy ship. Ōishi Matashichi writes how the Joint Committee on Atomic Energy issued a statement: 'the Japanese fishermen may have entered the danger zone for a purpose other than fishing and may have spied on the nuclear test' (35).

5 This is of course only *one* strand of the survivor subjectivity – memorialization, campaigns and active political protests against nuclearization have been a part and parcel of *hibakusha* and Japanese initiatives since the 1950s.

6 Lisa Lynch examining anti-reactor documentaries and nuclear energy activism has argued that these films and their protagonists do fit the model of deterritorialized subjects, whose awareness of the larger political and economic forces that shape lives across geographies and locales merges the local and the global (Lynch 2012).

3   The A-List

Atomic Scientists and Bombmakers

J. Robert Oppenheimer, often called the 'father of the atomic bomb', when recruiting people for Los Alamos and the bomb project, recorded how he argued the case with the scientists he wanted to join him:

> It was an unparalleled opportunity to bring to bear the basic know-ledge and art of science for the benefit of the country . . . This job, if it were achieved, would be a part of history . . . this sense of excite-ment, of devotion and of patriotism . . . prevailed. Most of those with whom I talked came to Los Alamos.
>
> (US Atomic Energy Commission 13)

A similar reason – patriotism – apparently drove the German scientist and Nobel Laureate, Warren Heisenberg, in his efforts to build the bomb for his country. In their biography of Enrico Fermi, Gino Segrè and Bettina Hoerlin observe about Heisenberg:

> Heisenberg was another matter. He had been a fervent adherent of the German youth movement, and his nationalism was deeply implanted. He was not seriously tempted to stay in the United States. His country needed him. His decision created scars and an abiding divide in the international scientific community.
>
> (155)

Patriotism and national pride, the biographies of the atomic scientists suggest, were a part of the intellectual and moral make-up of these brilliant physicists, chemists and engineers. The patriotic scientist is one of the most consistent *persona* that emerges from the biographies of the mid-twentieth-century atomic scientists. This creation of the scientific persona has a long history.

The 'experimental identity' of the scientist dominated the seventeenth century with its insistence on empiricism, and Robert Boyle would be a preeminent example of this kind of person (Shapin 1994). The scientist as a public figure emerged in Europe in the late nineteenth century (Källstrand 2020). Charles Darwin, Michael Faraday and Joseph

DOI: 10.4324/9781003254294-3

Banks were celebrity scientists, with a massive public appeal and cultural authority. Later, with the professionalization of science and new forms of funding and research especially in the American context, new modes of the scientific self emerged (Shapin 2008). In the twentieth century, but prefigured in the lives and careers of Lord Kelvin, Darwin and Faraday, the scientist as a tradable commodity is constructed through a blurring of their public and private lives, and within discourses of truth, reason and rationality, all located in the ideological climate of their era (Fahy and Lewenstein 2021). This celebratization and commodification of the scientist takes a giant leap forward in the wake of both, the Nobel and the appearance of television. The Nobel, long seen as an index of excellence in science, was based on specific and specialized research, although 'the Nobel prize becomes a medium, a way to publicise the science it awards' (Källstrand 61). The Nobel also endorses the (Western) cult of the genius through the mediatized publicizing of the Prize, the Prize-winner and the science: 'If the celebrity-scientist is a co-construction of the Nobel Foundation and TV media, the representation of Nobel Laureates as geniuses is a joint effort of the scientific community and TV media' (Ganetz 242). The idea of 'genius' is itself historical, serving as 'a categorical mode of assessing human ability and merit . . . [and] as a concept with specific definitions and resonances, genius has performed specific cultural work within each of the societies in which it has had a historical presence' (Chaplin and McMahon 1–2). The figure of the genius will resonate throughout the biographies of the atomic scientists, as we shall see.

Personae, including scientific personae, 'are the material forms of public selfhood' (Marshall et al. 290). 'Persona criticism' as Cheryl Walker (1991) and later commentators have called it, involves not only the life and mind of the brilliant/genius scientists but also the institutional and social contexts in which they practised their science. It focuses, as Walker puts it, 'on ideation, voice and sensibility, linked together by a connection to the author' (109). She qualifies the notion of the 'author' when she argues: 'authorship is multiple, involving culture, psyche and intertextuality' (109). Lorraine Daston and Otto Sibum offer the following account of personae:

> Intermediate between the individual biography and the social institution lies the persona: a cultural identity that simultaneously shapes the individual in body and mind and creates a collective with a shared and recognizable physiognomy. The bases for personae are diverse: a social role (e.g. the mother), a profession (the physician), an anti-profession (the *flâneur*), a calling (the priest) . . . Personae are creatures of historical circumstance; they emerge and disappear within specific contexts. A nascent persona indicates the creation of a new kind of individual, whose distinctive traits mark a recognized social species.
>
> (2–3)

Emphasizing the socio-cultural contexts, they add: 'To achieve a persona presupposes a certain degree of cultural recognition, as well as a group physiognomy that can be condensed into a type' (5). Kristi Niskanen et al. too underscore the social aspects of personae: 'they are collective entities, a kind of cultural and social repertoires on how to be a person of science' (1). The scientist 'deploys cultural vocabularies and repertoires that are mixed and assembled through cutting and pasting' (Bosch 13). Moreover, studying the scientific persona helps in 'revealing the intersections between the individual person, the cultural, scientific institution and the work that is conducted by the residents of this very institution for shaping a suit-able identity' (15). In the words of Herman Paul, 'scholarly personae [are] *models of scholarly selfhood*, or as models of abilities, attitudes, and dispositions that are regarded as crucial for the pursuit of scholarly study' (353, emphasis in original). Arguing that 'scholarly personae are more than momentary instances of scholarly self-fashioning', Paul writes:

> They are not identical to an individual's "performance" of scholarly identity, but provide the coordinates within which such performances are recognizable as acts of *scholarly* self-fashioning. They offer the templates that individuals appropriate in different ways, the themes on which they vary, the repertoires they enact, or the grammars they have to respect.
>
> (354, emphasis in original. Also Paul 2016)

This chapter examines the persona of the bombmakers as it emerges from the biographies. It unpacks the personal and professional, indi-vidual and institutional, 'scholarly' and/or 'scientific' identity of the men who built the bomb, as constructed by the biographies and related materials. As persona criticism/studies and the studies of the scientific-scholarly self cited above concur, social and institutional valorization as to what counts as the scholarly self, also provide repertoires of conduct – self-fashioning – for the scientist who seeks recognition and validation. The credibility of the scientist *qua* scientist depends to a large extent on the repertoires employed by the scientist and accepted by their dis-ciplines, the institutions and the world at large. My focus is not the self-fashioning by the scientists themselves but the *biographical* construction of the atomic scientist persona, with evidentiary and supplementary instances from any personal quotations and writings by them. This focus on the biographical construction of the scientific persona enables us to locate the atomic scientist within three crucial contexts – the Second World War, the Hiroshima–Nagasaki bombings and the subsequent cam-paign for delimitation and control of the nuclear arsenal – also traceable through these biographies. The biographical construction of the atomic scientist persona reveals multiple layers and nuances of what counted as a 'scientific self' and, frequently, the origins of this self in the scientist's childhood, modes of working and the milieu.

Many of the Manhattan Project scientists were celebrities by the time they came to Los Alamos/Chicago/Oak Ridge/Hanford – Nobel Laureates, and/or known as exceptionally bright physicists. Within their community, figures like Oppenheimer, albeit without a Nobel, had glittering reputations for being at the cutting edge of contemporary physics. Several were public figures already.[1]

The biographies themselves, published in the twenty-first century, trace the rise to celebritydom and the accumulation of reputational capital among scientific circles, of the atomic scientists. This chapter reads Jeremy Bernstein's *Oppenheimer: Portrait of an Enigma* (2004), Kai Bird and Martin Sherwin's *American Prometheus: The Triumph and Tragedy of J. Robert Oppenheimer* (2005), Giuseppe Bruzzaniti's *Enrico Fermi: The Obedient Genius* (2016), Ray Monk's *Inside the Center: The Life of J. Robert Oppenheimer* (2012), William Lanouette's *Genius in the Shadows: A Biography of Leo Szilard, the Man Behind the Bomb* (2013), Abraham Pais' *J. Robert Oppenheimer: A Life* (2006), Silvan S. Schweber's *Nuclear Forces: The Making of the Physicist Hans Bethe* (2012), Jost Lemmerich's *Science and Conscience: The Life of James Franck* (2011), Gino Segrè and Bettina Hoerlin's *The Pope of Physics: Enrico Fermi and the Birth of the Atomic Age* (2016), Charles Thorpe's *Oppenheimer: The Tragic Intellect* (2006), James Gleick's *Genius: The Life and Science of Richard Feynman* (1992), Matthew Shindell's *The Life and Science of Harold C. Urey* (2019), besides short obituary notices such as Hans Bethe's 'J. Robert Oppenheimer' (1997) and graphic science biographies like Jim Ottaviani et al.'s *Fallout: J. Robert Oppenheimer, Leo Szilard and the Political Science of the Atom Bomb* (2013) and Jim Ottaviani and Leland Myrick's *Feynman* (2011). The biographies that constitute this chapter invariably incorporate vast amounts of materials such as letters and anecdotal evidence from contemporaries.[2] Supplementary and supporting materials for the chapter come from works like Cynthia Kelly's *The Manhattan Project: The Birth of the Atom Bomb in the Eyes of Its Creators, Eyewitnesses, and Historians* (2007), Richard Rhodes' Pulitzer-winning history, *The Making of the Atomic Bomb* (1987) and Alice Kemball Smith's *A Peril and a Hope: The Scientists' Movement in America 1945-1947* (1971), that bring together interviews, reports (such as the famous Franck Report of 1945, arguing that the atomic bomb should *not* be used on Japan, or the 1950 Niels Bohr letter to the United Nations) and quotes by scientists. Public hearings such as the one employed to indict Oppenheimer in 1954 produced a wealth of documentation (collated in *In the Matter of J. Robert Oppenheimer: Texts of Principal Documents and Letters*, 1971) that also construct the atomic scientist in particular ways.

The biographies, expectedly, also provide a history of the science of the atom, as commentators have observed, indeed demanded, of the scientific biography. In an early essay on the genre, Thomas Hankins argued that the biography should 'reconstruct the intellectual make-up of his subject' (8). It must, he said, 'deal with the science itself' (8). Helge Kragh writes:

> the term "scientific biography" [is] taken to imply a biographical study with an emphasis on the science of the portrayed scientist, or at least one in which his contributions to science is dealt with in no less detail than his life and career.
>
> (270)

Others like Mott Greene argue that the scientific biography is about the 'subject that instantiates, in the life-work of the subject, the discovery or invention of some important principle in the life of the author' (733), but is ultimately about the scientist-as-hero. The life of the scientist intertwines with the life-work of the scientist *and* the history of the science of which h/she is a part.

The biographies, despite their diversity of approaches – a greater or lesser emphasis on the science, for example – develop the persona of the atomic scientists through a concentrated attention to the credibility and reputational capital of the scientist; the cultural scripts and repertoire of conduct that constitute the personal attributes of the individuals within particular contexts, mostly from the time of their childhood to later lives as scientists; and the development of specific virtues and moral conduct vis-à-vis the bomb and nuclear (dis)armament. The image of the scientific/scholarly self emerges through the coalescence of these discourses, cultural vocabularies and models.

The Credible Scientist and Reputational Capital

When Arthur Compton sought to establish a 'Metallurgical Laboratory' (Met Lab, as it was codenamed) at the University of Chicago, he approached the University's president, Robert M. Hutchins. Hutchins, writes Matthew Shindell, was

> receptive [to the idea] and came to the conclusion that he would be able to raise the necessary money for a new institute only if he had a stellar group of scientists with reputations like Urey's to help found it.
>
> (unpaginated)

In parallel, when looking for the Scientific Director to head Los Alamos, Leslie Groves sought a 'director [who] needed to have a stellar scientific reputation' (Segrè and Hoerlin 207). Clearly, the success of the project, whether in terms of funds raised or results achieved, depended on the choice and participation of scientists whose reputations were luminous. This discourse of reputation and its cognate, credibility, runs through all the biographies and has two principle strands: the renowned skills possessed by the scientist and the institutional-social position, power and role of the scientist.

Epistemic Virtues and Credibility

When Harold Urey took charge as the founding editor of the *Journal of Chemical Physics*, he had already published 'twenty papers or notes about atomic structure and experimental molecular band spectroscopy' and his 'status may have risen high enough to be entrusted with the editorship of the new journal' (Shindell, unpaginated). By the time Oppenheimer arrived in Göttingen for further studies, his papers 'represented small but important advances in quantum theory', and his reputation was already established (Bird and Sherwin 59). Bird and Sherwin add:

> Oppenheimer himself published seven papers out of Göttingen, a phenomenal output for a twenty-three-year-old graduate student.
>
> (63–4)

Now skills are oriented towards addressing delineated tasks and demonstrating the mastery over a certain technique (Paul 2014: 358). The emphasis on the professional skills of the scientists, demonstrated through their published work (and, in the case of Oppenheimer, the *rate* of publishing), from their very early years, in the biographies is coterminous with the emphasis on their reputational capital – all acquired by the time they arrived at or were co-opted into the Manhattan Project. That is, publications embody the peer acceptance of the skills of the scientist and constitute the making of their reputational capital. These are manifestations of discipline-specific skills – Oppenheimer's theoretical brilliance and poor math – that is central to the scientific-scholarly persona.[3]

The skills also embody what Herman Paul terms 'epistemic virtues' that they 'serve as an additional criterion for assessing the relative merits of the theories they propose' (2014: 349) and thus make up the scholarly persona. That is, the published work of the scientists already constructed them, in the eyes of their peers, as individuals with epistemic virtuousness which then produce the academic-scientific credibility in the field and an acceptance of their theories. This is borne out in an incident Bird and Sherwin record, involving Max Born, Oppenheimer's doctoral supervisor at Göttingen, to whom Oppenheimer had sent a five-page paper:

> Born was "horrified" . . . Born eventually lengthened the paper to thirty pages, padding it, Robert thought, with unnecessary or obvious theorems . . .
>
>   "On the Quantum Theory of Molecules" was published later that year. This joint paper containing the "Born-Oppenheimer approximation" – in reality, just the "Oppenheimer approximation" – is still

regarded as a significant breakthrough in using quantum mechanics to understand the behavior of molecules . . . The paper laid the foundation for developments more than seven decades later in high-energy physics.

(66)

Thus, the effort at writing, revising and publishing the paper is used by the biographers to reflect on the commitment of the scientists to the field but also on their credibility as contributors to the discipline. It is a reflection on Oppenheimer's 'dispositions', as Herman Paul terms them, that Born recognizes and legitimizes. As Ray Monk notes:

Born's respect for Oppenheimer was clear to everyone at Göttingen and seemed to elevate him above his fellow students.

(125)

Born's recognition of his student's epistemic virtue is reiterated in a letter he writes later to Oppenheimer, about the latter's parting gift of a classical text in physics:

This [book] has survived all upheavals: revolution, war, emigration and return, and I am glad that it is still in my library, for it represents very well your attitude to science which comprehends it as a part of the general intellectual development in the course of human history.

(Bird and Sherwin 67)

Born acknowledges that for Oppenheimer, science was a part of the general intellectual development of human history itself. It is not just the quality of his written/published work, but his sensitivity to what constitutes the best gift for a scientist, that lends Oppenheimer the aura of epistemic virtuousness in Bird and Sherwin's portrait. In Gleick's biography of Feynman, he writes:

Graduate students and instructors found themselves wandering over to the afternoon tea at Fine Hall with Feynman on their minds. They anticipated his bantering with [John] Tukey and the other mathematicians . . . Handed an idea, he always had a question that seemed to pierce toward the essence. Robert R. Wilson, an experimentalist who arrived at Princeton from the famous cauldron of Ernest Lawrence's Berkeley laboratory, talked casually with Feynman only a few times before making a mental note: Here is a great man.

The Feynman aura – as it had already become . . .

(1992, unpaginated)

The Gleick passage documents how both students and faculty were overawed by Feynman's brilliance, clearly pointing to the fact that it was his participation in the institutional life that enables the creation of the aura. From the life of Oppenheimer, we are given this example:

> One evening in March 1950, on the occasion of Einstein's seventy-first birthday, Oppenheimer walked him back to his house on Mercer Street. "You know," Einstein remarked, "when it's once been given to a man to do something sensible, afterward life is a little strange." More than most men ever could, Oppenheimer understood exactly what he meant.
>
> (Bird and Sherwin 382)

The Oppenheimer biographers imply the depth of understanding between two of arguably the greatest physicists of the era – and one widely acknowledged as a universal genius – when Oppenheimer is able to 'understand exactly' what Einstein meant (this is the Bird-Sherwin *construction* that Oppenheimer understood what Einstein meant, of course). This camaraderie but also the deeper insights into life, as portrayed by the biographers, construct the persona of the elder statesman (Einstein), the younger, brilliant discipline/student (Oppenheimer) where the insight into life by the former is transmitted to someone capable of receiving and understanding it. This acknowledgement is accorded by Einstein to Oppenheimer, even as Oppenheimer receives the older physicist's statement as both, due to him and as relevant to himself.

This kind of account is *not* of the Romantic solitary genius: rather it is an exceptional mind situated within a receptive setting of peers and betters that allows and creates the aura of the genius. The genius *pioneer*, the *discoverer*, the *founder* of an institution, note David Aubin and Charlotte Bigg in their account of the biographies of astrophysicists, are 'historiographically laden archetypes' that are

> central to the emergence and shaping of scientific disciplines in the late nineteenth century, as science took on the form of an increasingly collective endeavour, with groups of researchers identifying with well-defined procedures, conceptual frameworks, methods, approaches and instruments.
>
> (56–7)

They ask:

> Should we consider the individualistic types of the founder/explorer/discoverer, paradoxically endowed with all the more individuality as collective disciplines were subduing individual characteristics, as necessary counterparts, holding communities together by shared beliefs in them?
>
> (57)

Shared beliefs, exchange of knowledge and modes of inquiry consensually arrived at trouble the archetype of the solitary genius, as Aubin and Bigg indicate. Such a conceptualization enables us to see genius, as the examples from the biographies indicate, as the social valorization of a set of exceptional qualities and abilities: the college, the university and occasionally the state and its publics, who/which determine what counts as exceptional.

The credibility of the scientist was created, reinforced and celebrated through institutional mechanisms and the social/institutional role they played. The biographies document, both proleptically and analeptically, the professional affiliations of the scientists and education histories. In the process, something else also emerges.

Atomic Prosopography

Accounts of Oppenheimer's studies at Harvard, Cambridge and Göttingen are preliminaries to the account of his appointments to Caltech and Berkeley in Bird and Sherwin. When introducing the star cast of Los Alamos, the biographies invariably mention the Nobel Laureates present, and those who would win it in the future (Richard Feynman being one of the latter). Even when speaking of the opposition to the use of the atomic bomb, the biographies feel the need to highlight the reputations and credibility of the scientists by invoking their professional affiliation and achievements:

> Szilard organized a committee of like-minded souls, including most notably Glenn Seaborg, the discoverer of plutonium, under the chairmanship of James Franck, the Nobel laureate . . .
>
> (Monk 433)

Arthur Compton recalls his meeting with Oppenheimer:

> He [Oppenheimer] was a member of the colony of American students of James Franck and Max Born at Göttingen [and was] one of the very best interpreters of the mathematical theories to those of us who were working more directly with the experiments.
>
> (Monk 126)

This in itself is high praise, but Monk qualifies it with the following sentence of his own:

> Coming from the man [Arthur Compton] about to win the Nobel Prize, this is an extraordinary compliment to pay a twenty-two-year-old who had not yet completed his Ph.D. thesis.
>
> (126)

Reputational capital accrues to the demonstrated skills of the scientist in the form of such praise by peers, institutional privileges, rewards and appointments. In the process, the biographies of individual scientists also generate prosopographical narratives of what Alison Booth called 'inclusive circles of recognition' (https://onlinelibrary.wiley.com/doi/10.1002/9781118405376.wbevl260).

As the biographies strive to emphasize – and in contrast to Mott Greene's view of the scientific biography as a variant of the *bildungsroman* – the hero is heroic *because* he energizes a team of brilliant scientists to collectively work to find an answer. That is, in the place of a *bildungsroman*, we have a prosopographic narrative of scientific genius. The individual's scientific identity is located in a relational, contextual and collective setting of peers and institutions, even (or especially) when the individual scientist is a public figure (like Oppenheimer or Einstein). The construction of the scientific identity – or personas – of these individuals in biographies 'are embedded with strategy and intention and that personas are the material forms of public selfhood, even when entirely composed of digital objects, network connections and mediated expressions' (Marshall et al. 290). In the nuclear biography, the public selfhood is the effect, of course, of the network connections but also of the reputational economies that emerge from these connections and mediations. These latter are made visible in the prosopographic aspects of the auto/biographies.

Helen Kragh points to this trend in scientific biographies where 'biography is skillfully blended with a prosopographical study of . . . [a] network of assistants, astronomers, clients and patrons' (273). Such biographies focus on the environment of the scientist and the 'social forces' shaping the individual's science (274). Steven Shapin and Arnold Thackray argue that prosopography is a mode of historical analysis that suits the history of science because it enables the study of a group of actors, especially because (in the nineteenth century in England), it affirmed a new social and economic order and because participation in a scientific culture distinguished individuals from those who had no cultural pursuits (1974). (The sharing of knowledge that the network demanded and made possible was integral to the group identity and the individual scientist, and the insistence of openness among scientists in the Manhattan Project – much to the discomfiture of Leslie Groves, who wanted compartmentalization – stems from this feature of knowledge-making in scientific societies and groups.) Early commentators on the genre noted that while the 'researchers establish the group to be studied, such as a profession, society or college', but for those groups without 'formal membership qualifications', such a study becomes more complicated (Sturges 212. See also Verboven et al.). In the biographies of the atomic scientists, the prosopographical component recognizes that there was no 'atomic scientists club' or coherent group: they had been bought together through contextual proximity, reputational capital and intellectual camaraderie. Therefore, we find the narratives emphasizing the 'who knew whom', 'who was whose student' and 'whose work was familiar and/or useful to the Project'. We see this borne out as late as 1965. When Oppenheimer spoke of Einstein in 1965, he first emphasized the latter's 'extraordinary originality' (36), and then went on state: 'Einstein brought to the work of originality deep elements of tradition. It is only possible to discover in part how he came by it, by following his reading, his friendships . . . ' (36). Oppenheimer then lists how the work of de Broglie, Max Planck and Maxwell among scientists helped Einstein think, thus outlining the contextual and relational situatedness of Einstein's work and greatness (37). Leslie Groves, in his account of the Manhattan Project, states in his Foreword to *Now It Can Be Told*:

> No single stroke of genius delivered up the finished product. Rather, its present state of development derives from the labors of many individuals from many countries, operating in many fields of endeavor, over a span of many years.
>
> (unpaginated)

Oppenheimer biographies of course underline the context – wartime – and the genesis of the Manhattan Project in secrecy, national fervour, state support, among others. But they also invariably point to the *kind* of scientists who, despite the secrecy (then) around the work they did, constituted the network that collaborated with Oppenheimer.[4]

Then, all biographies emphasize that the Manhattan Project was peopled by the best theoretical and experimental physicists in the country. Oppenheimer, we are told, called the group his 'luminaries' (Bird and Sherwin 181). Several of the 'luminaries' were united by a common theme: they were immigrants to the USA – like Hans Bethe, Enrico Fermi, Eugene Wigner, Leo Szilard and Niels Bohr, all led by Oppenheimer whose father, Julius, had migrated from Germany in the 1880s. Then, many of those Oppenheimer recruited, such as Robert Serber (famous for the introductory lectures on the Project, delivered at Los Alamos, which would then become *The Los Alamos Primer*), were his former students. There are other examples too. For instance, Joseph Rotblat, who would quit Los Alamos because he discovered that Germany would never be able to make an atomic bomb in time to win the war – therefore, in his view, the USA's building of the atomic bomb was not *necessary* – and went on to win the Nobel Prize for Peace for his anti-nuclear efforts, had been picked by James Chadwick, the discoverer of the neutron, as a PhD student. Chadwick was also part of the Los Alamos team, from the UK's side. Isidor Rabi, Nobel winner in 1944, consultant on the Manhattan Project, was a postdoctoral student of Albert Sommerfeld at Munich, with Rudolf Peierls and Hans Bethe. In Copenhagen, Rabi and Wolfgang Pauli were associated with Niels Bohr.[5]

Biographers offering the prosopographic narrative of Los Alamos would, as noted, proleptically and analeptically, highlight the illustrious nature of the team headed by Oppenheimer. Jeremy Bernstein puts it most succinctly:

> Never before, or since, has such a collection of scientific talent been assembled to carry out one task. I made a list of the people there who either had the Nobel Prize before they came or received it after they left. I am not sure my list is complete. Fermi and Bohr had Nobel Prizes when they came. Bethe and Rabi got theirs later. Rabi was not on the staff at Los Alamos, but he acted, on his frequent visits, as a sounding board for Oppenheimer . . . The experimental physicists Norman Ramsey and Emilio Segre won the prize, as did the theorists Felix Bloch, Richard Feynman, and Bohr's son Aage, as well as the experimentalists Edwin McMillan, Owen Chamberlin, Fred Reines, and Luis Alvarez. Two of the more interesting cases were Val Fitch and Joseph Rotblat . . . He [Fitch] was a professor at Princeton in 1980 when he won his Nobel Prize for Physics for his work on elementary particles . . . In 1995 [Rotblat] won the Nobel Peace Prize.
>
> (2004: unpaginated)

Priscilla McMillan refers to the 'constellation of geniuses' (85).

This 'inclusive circle of recognition' as prosopography is also visible in the form of the photographs of the biographical subject, of which I focus on two kinds.

Over and above the (routine) photographs of their childhood and families, the biographies invariably include photographs of the subject with his peers. The Lanouette biography of Leo Szilard carries photographs of Szilard with Ernest O. Lawrence, Eugene Wigner, members of the Chicago group that created the first nuclear chain reaction, and others. Bernstein's biography has photographs of Oppenheimer with Ernest O. Lawrence, Edward Teller (a later photograph, from 1963, well after Teller had testified against Oppenheimer at the 1954 hearing). The Urey biography has photographs of Urey working in his laboratory at the University of Chicago. Gleick's account of Feynman includes images from Los Alamos, and Feynman with the team, and later ones with Abraham Pais, and others. Bird-Sherwin include photographs of Oppenheimer with Lawrence, Fermi, John von Neumann, Hans Bethe and Niels Bohr. In two cases, that of Urey's and Szilard's biographies, images of the film and filmmaking about them have been included as well: a biopic within a biography, so to speak. The printed version in the *Bulletin of Atomic Scientists* of Oppenheimer's 1965 speech on Einstein includes a photograph of them together (37).

Now the family portrait/photograph, argues Marianne Hirsch, generates an 'affiliative look' and a 'familial gaze', and we respond to 'dominant mythologies of family life' in which the family members in the photograph 'define themselves in relation to each other in the roles they occupy as mother, father, daughter, son, husband, or lover' (Hirsch vi).[6] But in biographies, such family pictures are 'between the family album and the public memorial' (vii) merging the two facets of the scientist's life, and almost rendering the genius a 'family man' (although most biographies document how the brilliant scientist was often wanting in his role as father/husband). But what of the photographs of Szilard, Oppenheimer, Feynman and others with fellow-scientists, colleagues and students? I propose that the affiliative look here is extended to beyond the family so that we encounter the *fraternity* of scientists – thereby making the photographs a prosopographic narrative.

One further point that shows the embedding of the individual's life in both, the prosopographic narrative and the history/story of the science. When images of the Trinity test, the laboratories and the instruments are placed within the individual biographies, they frame the individual's life within the life/story of those non-living actants, bringing them all into a network of actors that worked together. We recognize that the biography is the adaptation of the work in those laboratories, among those machines, for our consumption because the scientific elements incorporated as art-objects (photographs) are a part of this biographical narrative. We are *not* allowed to lose sight of the science or its sites/sights in favour of the individual story. Images of the laboratories and sites function as 'textual transferals' (Andrews 370). Textual transferals enable us to move from (i) the history of the science before and after the bomb and (ii) different domains within the science of the bomb (the chemists, the physicists, the mathematicians and the statesmen) *to* the life of the scientist and *vice versa*. That is, the photographic records of the laboratories and instruments are a conduit for the reader to move from one textual history (of the science) to another (the scientist). In the process, of course, the prosopographic narrative is reinforced.

In the genre of graphic scientific biographies – of Feynman, Oppenheimer – there is an additional layer to the biographical/prosopographical narrative made possible by the medium. Candida Rifkin argues in her reading of Jonathan Fetter-Vorm's *Trinity: A Graphic History of the First Atomic Bomb* (2012) and scientific biographies such as Jim Ottaviani et al.'s *Fallout: J. Robert Oppenheimer, Leo Szilard and the Political Science of the Atom Bomb* (2001):

> these works construct the figure of the scientist as a visual icon who is also a seeing subject. I propose that we consider scientific graphic biography as a specific genre that installs a biographical eye (as opposed to the auto/biographical I) to convey the complex relationship between empirical knowledge and affective experience that shapes lives caught between science and politics.
>
> (Rifkind 2)

Rifkind's focus is on the portrait of Oppenheimer, on his 'mythologization' (7). Along similar lines as Rifkind, I have argued that *Trinity* iconizes science as much as it iconizes the scientist (Nayar 2018).

Additional evidence for the prosopographic aspect of the auto/biography also appears in the hearing of 1954 and Oppenheimer's letter. Oppenheimer strives to downplay the lone hero image in his letter to K.D. Nichols:

> At the time, it was hard for us in Los Alamos not to share that satisfaction, and hard for me not to accept the conclusion that I had managed the enterprise well and played a key part in its success. But it needs to be stated that many others contributed the decisive ideas and carried out the work which led to this success and that my role was that of understanding, encouraging, suggesting and deciding. It was the very opposite of a one-man show . . .
>
> (14)

And later:

> I had become widely regarded as a principal author or inventor of the atomic bomb, more widely, I knew, than was warranted.
>
> (16)

Oppenheimer focuses here on the collective work of the scientists, spread across Oak Ridge, Hanford, Chicago and Los Alamos.

Now, while the prosopographic narrative explored 'the details about individuals in aggregate and analysing the characteristics that provide a broader understanding of society' (Oldfield 1865), the biographies also insist on portraying the uniqueness of individual scientists. That is, having documented the networks of recognition and the group/peer/institutional acknowledgement of the individual's epistemic virtues, the next (but not necessarily sequential) step in the construction of the persona of the atomic scientist is an overwhelming detailing of the *personal* attributes of the atomic scientist.

The Scientific Self

James Gleick's biography of Feynman comments: '[it] seemed to colleagues that some of his computation was a matter of conscious reputation building' (Gleick, unpaginated). Gleick also documents the Freeman Dyson disagreement with Oppenheimer over Feynman's theories (Gleick calls it 'war', unpaginated). Gleick writes:

> Oppenheimer did set up a series of forums to let Dyson make his case. They became an occasion. Bethe came down from New York to listen and lend moral support . . . In the end Bethe turned Oppenheimer around. He cast his vote explicitly with the Feynman theory and let the audience know that he felt Dyson had more to say. He took

Oppenheimer aside privately, and the mood shifted. By January, the war had been won.

(unpaginated)

In Ottaviani and Myrick's biography of Feynman, we are shown Hans Bethe and Feynman engaged in furious arguments, calling each other 'crazy' and dismissing the other's thoughts, ideas and calculations as 'wrong' (64). Argumentation and disagreement, the biographies suggest, are constitutive of the scientific self.

The scientific self, as the biographies map it, has three personal attributes: genius, charismatic authority and patriotism.

The Genius and the Atomic Hero

He [Oppenheimer] had been one of the most famous men in the world. One of the most admired, quoted, photographed, consulted, glorified, well-nigh deified as the fabulous and fascinating archetype of a brand new kind of hero, the hero of science and intellect, originator and living symbol of the new atomic age.

Robert Coughlan in *Life*, cited in Bird and Sherwin (556)

Exploring the scientific auto/biographies of the French scientists in the Enlightenment, Dorinda Outram observes 'a focus on episodes of commitment within the autobiographies themselves' (92). Outram writes:

Moments of epiphany, of absorption in Nature, in scientific autobiography have the same role as conversion moments in spiritual autobiography: they resolve the antagonism of the self and the world.

(93)

In his biography of Richard Feynman, James Gleick writes:

dreamily wondering how to harness atomic power for rockets, he worked out a nuclear reactor thrust motor, not quite practical but still plausible enough to be seized by the government, patented, and immediately buried under an official secrecy order.

(unpaginated)

From his childhood, Robert Oppenheimer was a dreamer:

"He was a dreamer," said Babette Oppenheimer, "and not interested in the rough-and-tumble life of his age group . . . he was often teased and ridiculed for not being like other fellows." As he grew older, even his mother on occasion worried about her son's "limited interest" in play and children his own age."

(Bird and Sherwin 15)

Oppenheimer's life in the 1930s, notes Ray Monk, was marked by an 'utter absorption in physics at this time' to the exclusion of any awareness of the world around him:

> If he showed some interest in things of beauty, such as the literature he read, he showed almost none in the social and political upheavals that were happening at that time.
>
> (27)

About Leo Szilard's childhood, William Lanouette writes:

> He usually ignored the mildly patriotic stories and turned right to the instructions and drawings for do-it-yourself household gadgets. With his clumsy hands, Leo seldom actually built these items; that was Bela's duty. More often Leo treated the plans as topics for intriguing daydreams . . . Leo let one thought run to another, creating landscapes of gadgets in his busy mind.
>
> (58)

Growing up and attending the university, he found the lectures and the explications boring:

> Szilard soon found himself impatient with the practical subjects taught at the institute and began daydreaming about physics.
>
> (Lanouette 58)

This image of the scientist as dreamer, self-absorbed, lost-to-the-world and unfit for/indifferent to the quotidian, as seen in the above descriptions, is a commonplace in the biographies. It is akin to what Geoffrey Cantor identified as the 'Romantic' account of the scientist with its stereotype of the unworldly genius (1996). The scientist is depicted as experiencing and exhibiting moments of (intense) absorption, epiphanies and demonstrating episodes of commitment, right from childhood.

The intensity-theme, which dovetails into the (Romantic) image of the self-absorbed genius, runs through the nuclear biographies. Even the conversations around the scientists in the Manhattan Project, note Segrè and Hoerlin, were intense: 'they spoke to one another softly, but intensely'. Elsewhere, they note Fermi's commitment and intense focus on the problem at hand:

> Fermi's stance was compatible with his extraordinary ability and predilection to compartmentalize physics, devoid of any distractions – political or otherwise. His devotion to science was unwavering, and probing its complexities was inevitable.
>
> (252)

There would be no distractions for the genius: the focus and mental energies coalesce around the physics problem at hand and nothing else intervenes. While this is not self-absorption, it is a particular image of the absorbed individual.

The genius scientist was a hero and a relentless quester after a secret. As early as 1927, when still a student, Oppenheimer's teacher saw in him an unraveller of secrets:

> [Professor Edwin Kemble observed that] his former student [Oppenheimer] seemed steeped in the excitement of unraveling the mysteries of quantum mechanics.
>
> (Bird and Sherwin 61)

But Oppenheimer was a discoverer and unraveller of secrets in other ways too. Bird and Sherwin, in order to draw attention to the original and unconventional thinking in Oppenheimer, tell us that his theoretical insights drove the experimentalists to discover things, where Oppenheimer *anticipated* the experimental results.

> It took someone like Oppenheimer to push [Paul] Dirac into predicting the existence of antimatter. This was Oppenheimer's penchant for original thinking at its best. In 1932 the experimental physicist Carl Anderson proved the existence of the positron, the positively charged antimatter counterpart to the electron. Anderson's discovery came fully two years after Oppenheimer's calculations suggested its theoretical existence.
>
> (Bird and Sherwin 87)

The quester here himself does not do the experiments that generate the results but provides the route for the experimentalist to march on his quest.

The scientist had to unravel the secret for, as David Kaiser has argued, the discourse around nuclear energy congealed around the idea of a secret, a formula or method:

> Beginning late in 1948 and accelerating through the mid-1950s, the weight of discussion among politicians and journalists shifted, focusing instead on textual and theoretical "information" as the essential "secret" of the atomic bomb, rather than experimental skill or industrial capacity. Many now claimed that specific, esoteric formulas – the x's and y's of theoretical physics – contained the true secrets of the atomic bomb.
>
> (34)

While Kaiser focuses on the early Cold War years, the origins of the discourse may have been in the run-up to the atom bomb itself. The genius scientist-as-uncoverer of secrets is a Romantic conceptualization of the scientist, notes Geoffrey Cantor (172).

Many present, as expected of the romantic strain in biographies of scientists (Cantor 1996), the scientist as endowed with a superior *imagination*. Ronald Clark's account of Einstein is cited by William Lanouette:

> Einstein's God thus stood for an orderly system obeying rules which could be discovered by those who had the courage, the imagination, and the persistence to go on searching for them.
>
> (88)

Wolfgang Pauli's view of Oppenheimer, likewise, commented on the latter's powerful imagination:

> His strength is that he has many and good ideas, and has much imagination.
>
> (Monk 160)

One of Oppenheimer's papers is described thus:

> Even to have raised the question of indefinite gravitational collapse required impressive boldness and imagination.
>
> (Monk 249)

Resonating with the Romantic ideal of the link between childhood and the imagination, the Szilard biography cites his own view: that his childhood continued to be a part of his adult life, especially as it impelled his scientific imagination (24).

In line with this Romantic conceptualization of the genius-scientist, by titling their biography *American 'Prometheus'* Bird and Sherwin are clearly invoking the myth of the unraveller of nature's secrets itself.[7] And this invocation deserves some attention, particularly when iterated in other biographies.

Fetter-Vorm's visual representation of the atomic bomb and the Manhattan Project begins not with science but with the scientist, Oppenheimer, who, driving towards the test site (Trinity), enlightens his vehicle's driver on an ancient myth: of Prometheus. The image of the tower in which the atom bomb is placed is an image, one could say, of contemporary science. But this is preceded by the image of Zeus cursing Prometheus who, chained to the mountain, is having his liver eaten by a bird. The entire top half of this page (3) is taken up by the visual representation of the myth. Following this, we see Oppenheimer toss out the remnants of his cigarette and saying: 'he [Prometheus] gave humans knowledge for which we weren't ready' (3). In the last panel on this page, we see the tower that science built, and the text box that appears to label the tower states: 'another ancient secret was about to be revealed' (3). Several things arrest us on this page's visual dynamics. First, it clearly aligns Oppenheimer with the mythic Prometheus and anticipates via the myth, the punishment that awaits the unraveller of the (forbidden) knowledge of the atom bomb: ostracized, ridiculed and even rejected by the power (US government/Zeus) for his science, a science for which humanity is not, ostensibly, ready. The juxtaposition of myth and science on the same page therefore forces us into the proleptic, or oracular, reading of the *science* and the *scientist*: this is the fate that awaits him, Oppenheimer. Second, it enlists antiquity and mystery in describing the present-day science: the 'ancient secret' is at once fire and the atom bomb. Through this, Fetter-Vorm mythologizes and romanticizes science itself: is a field of inquiry that unpacks ancient mysteries. It is not 'new' science, but a science that enables us today to finally reveal the world's oldest mechanisms of power and energy. Third, the two segments – the Prometheus myth and the atomic tower – are divided in terms of the page's layout the image of the vehicle in which Oppenheimer is travelling towards Trinity. The roof of the jeep appears to be either aflame or covered in cloud, roiling down from the abode of the Gods. The Zeus location and that of Prometheus has clouds as their border, but this seems to grow tongues of flame th extend over Oppenheimer's vehicle. The image is striking because it offers us a *mediating* link between the ancient myth and contemporary science fire. Prometheus and Oppenheimer are both drawn on the same side of the page, in a nearly straight line, positioning Oppenheimer as a figurative descendant of Prometheus.

By locating Prometheus and then Oppenheimer at the centre of the image and its action, Fetter-Vorm clearly aligns the two under the category of historically heroic figures. One final point in this representation of myth and science has to do with the flung-out cigarette. As the cigarette spins away into the clear white space of the New Mexican desert, we see the tip is still aglow. I see this innocuous object as central to the page's rhetoric. Examining the role of objects in visual representations, Joanna Woodall (2012) has argued that even as the human protagonists frame these objects, the objects frame the human interactions. The seemingly unimportant object in the frame, or even at the margins, offers us a way of interpreting the scene unfolding (Woodall). If the tongues of flame on the roof of Oppenheimer's car serve as a mediating link, the cigarette with its glowing tip points to the *human production and consumption* of the fire/fiery object, Fire, mediates the boundaries between humans and Gods, and now between (perhaps) life and death. Fire is what Prometheus secures; it is what Oppenheimer consumes in the form of the cigarette, and fire will be the possible outcome of the object in the tower, an object 'with the potential to burn as bright as a star'. The glowing cigarette tip enables us to see that at the centre of the ancient myth and modern science is a simple process: burning. The visuality of scientificity, then, is inaugurated not with the science *qua* science but with a linkage between an ancient myth and modern science, a linkage symbolized in the combustion process that is the cigarette: fire.

If Fetter-Vorm generates the Romantic biography of Oppenheimer as Prometheus, other biographies do it in a slightly different fashion – by implying that Oppenheimer foresaw the nuclear arms race. George Keenan would put it thus at the funeral oration for Oppenheimer:

> No one ever saw more clearly the dangers arising for humanity from this mounting disparity. This anxiety never shook his faith in the value of the search for truth in all its forms, scientific and humane.
>
> (Cited in Bird and Sherwin 5)

When Keenan speaks of a quest for truth in 'all its forms', it signals Oppenheimer's driving force, indeed his very reason for existence and life as a physicist. When he implies Oppenheimer's foresight, he bestows a Romantic seer role upon the scientist. This Keenan metaphor is visualized by Fetter-Vorm as well, this time in his drawing of the Trinity explosion itself.

Oppenheimer shades his eyes not to the blast at Trinity but to the future itself. We are told at the beginning of the narrative: 'what this bomb needed was a visionary' (21). Fetter-Vorm aligns this *vision*ary behind the bomb, with the visionary who *sees the future* of mankind and the atomic age. When Oppenheimer recites the key phrase of having become death and the destroyer of worlds, as he shields his eyes to the blast, the representation is no longer merely about the dazzling brightness of the scientific achievement, and the brilliance of the values that created the bomb or the military value of Oppenheimer's creation. Rather the image of the *visionary* Oppenheimer guarding his eyes proposes an entirely different value to the science – destruction and death. This is the future Oppenheimer-the-visionary is reluctant to see with his eyes. This interpretation of Fetter-Vorm's visual representation of the moment of scientific and personal triumph is invited by the next image of the whitened space of the pages, with the twisted stumps of the test tower and the text box which imbricates science with politics. The quest for knowledge has produced a scene unbearable to see, suggests the portrait of the bombmaker.

And yet, the scientist's personal quest for truth, notes Mott Green (2007), grounds the very structure of the scientific biography. The hero must overcome obstacles, travel to another place, find a magical element and eventually solve the problem, generating a satisfactory resolution. As Geoffrey Cantor puts it in his study of the public images of Michael Faraday:

> For the Romantics genius is, of necessity, not prosaic but possesses some extraordinary, even mystical, power which enables the natural philosopher to divine nature and discover its secrets.
>
> (179–80)

Such a romanticized portrait of the genius-scientist, tortured, fragmented and luminous is the traditional Oppenheimer portrait. And if Bird-Sherwin invoke the Promethean myth, other metaphors and symbols accrue to Oppenheimer as well.

For instance, Silvan Schweber modifies the trope and the myth in his account of Oppenheimer. Schweber describes him as 'Protean' for his remarkable ability to adapt from his physicist role to that of an administrator. Schweber then argues:

> Oppenheimer made many things of himself: creative physicist and influential teacher in an era that revolutionized the physical sciences, charismatic administrator of a wartime project that altered the course of world history, prominent adviser to the highest echelon of American policy makers in the postwar period. In each of these roles he became the personification of what others should aspire to. Yet for all these accomplishments he could not fashion a sense of identity for himself.
>
> (220)

This was so because, in Schweber's reading,

> he [Oppenheimer] found it difficult to integrate these disparate roles – perhaps because he found it difficult to conceive an overall creative vision for himself or to devise a compelling objective for the community he belonged to if one had not been formulated at the time he assumed its leadership; perhaps also because each of these activities had also been connected with a deep crisis – a deep rupture.
>
> (220–21)

If Bird and Sherwin portray Oppenheimer as a genius punished for stealing a secret from Nature, Schweber chooses to depict a genius whose personality is tormented and fractured. Isidor Rabi would endorse this view:

> a man who was put together of many bright shining splinters . . . [who] never got to be an integrated personality.
>
> (Cited in Monk 3)

Lindsey Banco points to these depictions of Oppenheimer when she writes:

> The primary ambivalence these biographies engage with (but rarely resolve) is the supposed antipathy between the hyperrational quest for knowledge attached to Oppenheimer's identity as a scientist and the many not unproblematic signals encouraging us to read him as an American version of the Romantic artist tapping intuitively or emotionally into unearthly and sublimely incomprehensible knowledge.

The scientific portrait of Oppenheimer frequently turns on the idea that the rational process of acquiring knowledge can be metaphorized in the scientist's privileged position surveying a vast stretch of landscape and looking at it with a penetrative gaze. The second type of portrait, the one informed by the Romantic figure of the artist, relies on emphasizing emotional responses to the landscape and using metaphors of gnosticism to position Oppenheimer in the desert as a determinative, shaping force.

(494)

An element of magic, sublime ability and emotional-intellectual fragmentation mark these accounts, all of which contribute to the genius persona of Oppenheimer in particular. Even at the inquiry, the portrait of an enigmatic genius emerges in the Lloyd Garrison summation:

You have in Dr Oppenheimer an extraordinary individual, a very complicated man, a man that takes a great deal of knowing, a gifted man beyond what nature can ordinarily do more than once in a very great while. Like all gifted men, unique, sole, not conventional, not quite like anybody else that ever was or ever will be.

(US Atomic Energy Commission 990)

But for the nuclear scientist at Los Alamos and its related sites in Chicago, Hanford and Oak Ridge, the magical element was more or less something they *knew*: uranium and plutonium, thus their enchantment with a 'nature's secret' was an informed one, for the techniques and theories to unravel this secret to the core were already in place (although they were not assured of success). Their struggle was to acquire enough of the materials in pure form and device ordnance that would implode to generate the maximum effect. The travels were, of course, to and fro Los Alamos. The larger problem to be addressed was Nazi Germany and *its* quest for atomic power.

The biographies document the obstacles and problems in considerable detail: from making the New Mexico site suitable for the laboratory and scientists' residence, to shipping the materials needed; from designing the theory of the 'gadget' (the nickname for the bomb) to the structure of the ordnance; from simultaneously sharing and compartmentalizing information about the bombmaking process; to the decision/support for the bombing of Japanese cities. In the case of the last, the obstacle was mainly the tension between the success of the science of the bomb, its experimental results now clear, and the terrible consequences of deploying it.

Nuclear Patriotism

Genius aside, a key personal attribute and contributor to the scientific self as constructed in the biographies is the patriotism of the scientist.

The Manhattan Project was a war initiative and effort and as such the rhetoric of national security and patriotism was commonplace. Biographies of the nuclear scientists repeatedly draw attention to the patriotism and national fervour of the scientist. Patriotism is depicted as a key mode of functioning, an attribute and a code of conduct here. The biographies capture, I suggest, not the science alone but the way in which science works (Hankins 1979).[8] Nuclear patriotism is not an emotional, hyperventilating or jingoistic offensive but presented as a more rational kind of nationalist sentiment in the scientific biographies.

The discourse of nuclear patriotism underlies and sometimes overlays the biographies' account of the lives of the scientists, their scientific work and the organization of this work – between the scientific and the military. The science, in other words, is subject to, according to the biographies, a code of conduct influenced by the war. The scientist, from the repertoire of conduct available to him, chooses national needs and patriotism.

In his well-known biography of Oppenheimer, Charles Thorpe writes:

> I argue that (paradoxically, in light of the security hearing) Oppenheimer in significant ways accommodated himself to and internalized the culture and mentality of the national-security state.
>
> (xv)

Oppenheimer's own account of the patriotism that drove the scientists to join the Manhattan Project has already been cited at the beginning of the chapter. Robert Marshak, a fellow scientist at Los Alamos, attributed the long hours spent working to 'curiosity and zeal . . . [and] an inspiring patriotism' (cited in Thorpe 133). Ray Monk, tracing Oppenheimer's education and development as a scientist, emphasizes Oppenheimer's patriotism which merges with a burning scientific ambition:

> combined with his fervent patriotism, this [the determination to be at the center of scientific discovery] drove Oppenheimer to make America the world center of advances in physics.
>
> (13)

Later, Monk would align the developments in physics within the USA with the arrival of numerous Jewish migrants from Europe and joining various American universities. Monk writes:

> Along with (to mention only the most prominent) Einstein at Princeton, Hans Bethe at Cornell and James Franck at Johns Hopkins, [Felix] Bloch thus became part of the extraordinary enrichment of American physics that was brought about through the absorption of Jewish émigrés. Indeed, within a few years the United States had replaced Germany as the world's leading center for the study of physics, partly because many of the people who had made Germany

preeminent in the field were now working in American universities. As the relentlessly patriotic Oppenheimer was quick to point out, these refugees would not have had the impact they did had there not been "a rather sturdy indigenous effort in physics" . . .

(214)

Further evidence of Oppenheimer's nuclear patriotism is visible in accounts of his keenness to place American physics on a high pedestal, which simultaneously would contribute to the war effort:

In the 1930s he had set out to build an *American* school of theoretical physics that would enable the U.S. to replace Germany as the leading center for research in that area; now he had a chance to lead a project that would not only demonstrate the superiority of American physics, but would also, in so doing, equip the U.S. with a weapon that would enable it to win the war against Germany.

(Monk 336, emphasis in original)

When he took charge at Los Alamos, Oppenheimer got himself an officer's uniform. Kai Bird and Martin Sherwin write in *American Prometheus*:

After the physical, Oppenheimer had an officer's uniform tailored for him. His motivations were complex. Perhaps donning a colonel's uniform was a visible sign of acceptance important to a man who was self-conscious about his Jewish heritage.

(210)

Bird and Sherwin even title a chapter 'He'd Become Very Patriotic'.
    Nobel Laureate Hans Bethe, at one point Oppenheimer's colleague at Los Alamos, would summarize Oppenheimer's *American* role as follows:

Oppenheimer created the greatest school of theoretical physics that the United States has ever known. Before him, theoretical physics in America was a fairly modest enterprise, although there were a few outstanding representatives.

(177)

Oppenheimer, says Bethe, 'in 1942 . . . felt the deep urge to contribute to the American war effort . . . [he] had the great desire to identify with the U.S. war effort, and was quite ready to accept a commission' (187–8).
    Abraham Pais cites Oppenheimer's friend, Charles Wyzanski, who, coming to know of the impending hearing in 1954, wrote to Oppenheimer:

[M]y proposal is that you forthwith publish in a medium of wide circulation (say the *New York Times,* or perhaps *Life,* or *The Atlantic)* a more than candid, a philosophical autobiographical account of your intellectual curiosity, your detailed conduct, your personal and family relations, your and their experiences, your record of patriotic contribution, and your understanding loyalty. This would be more than a psychoanalytic performance. Its literary equivalent might be Montaigne's Essays. But it would be something far greater because it would be an unprecedented contribution to the political education of this country. The true "witness" this nation needs is not a recusant Communist but an independent citizen who can explain what it was like in the '30's and '40s to be a man of character and adventurousness seeking to learn the truth about one of the undeniably major forces of our time, and simultaneously to advance the welfare of mankind.

(264)

At his funeral service, George Keenan, former ambassador and advocate of the postwar containment policy against the Soviet Union, said of Oppenheimer:

It was the interests of mankind that he had in mind here; but it was as an American, and through the medium of this national community to which he belonged, that he saw his greatest possibilities for pursuing these aspirations.

(Bird and Sherwin 5)

Nuclear patriotism, as one can think of the above context for Oppenheimer's work as a scientist, aligns the work *in* science with the national(ist) work *of* science. The scientist-as-patriot is represented as exemplifying a rational nationalism and patriotism (Lindsey Michael Banco makes this point about the Bird-Sherwin biography 499). The biographies do so by embedding the discourse of patriotism within a discourse of national needs *and* science's role in meeting those needs.

The biographies, then, portray Oppenheimer-the-scientist as committed to the project of making the bomb because it was not just a scientific exercise but a national-security one. Writing about scientific biographies, Mott Greene argues:

What biography also accomplishes . . . is specific knowledge of how cultural movements and political or scientific developments come together in a given time and place. It allows this by recreating the conjunction of these entities, motives, ideas, events, and perceptions in the life and mind of a single subject.

(729)

Greene further proposes that:

the combination of consistency and autonomy must lead to a view of the life of the subject that instantiates, in the life-work of the subject, the discovery or invention of some important principle in the life of the culture, and which can function as the origin story of some belief, idea, or convention.

(733)

The Oppenheimer biographies thus uniformly project his nationalism in the war years.

Other nuclear scientists like Warren Heisenberg, biographers note, were similarly wedded to a national destiny and a national cause. In their biography of Enrico Fermi, Gino Segrè and Bettina Hoerlin's observation about Heisenberg's 'fervent' nationalism (155) has already been cited above. Fermi himself, being an immigrant, raised some doubts in the minds of the American policy makers, as the biographers note

> [Arthur Holly] Compton had been very cautious in involving Fermi because of security concerns. Fermi was not an American citizen. But Samuel Allison, Compton's Chicago colleague and protégé, told him that Fermi was absolutely the man he needed to speak to first. When Compton had asked Allison who could give him a reliable estimate of how much U-235 would be needed for a bomb, Allison's answer had come quickly, "No one can answer that question as well as Enrico Fermi."
>
> (Segrè and Hoerlin 171–2)

As the biographies capture, the scientists were embedded in a wartime discourse of national security. Even into the mid-1950s, as Ralph Gabriel described it in an essay in 1957 (just three years after the Oppenheimer hearing), 'during the brief period in which the United States had a monopoly of atomic power the American people enjoyed an invincible security' (542).

The emphasis on the scientist-as-patriot positions the American scientific community and the science itself as a technology that needs to be hierarchically organized. The same technology if developed by the Germans would be catastrophic for the *world* and America in particular, and hence it is better that the science of the atom bomb be developed first, and faster, in America. Segrè and Hoerlin put it this way:

> Fears that Germany would develop the bomb before the Allies succeeded in doing so were particularly pervasive among Los Alamos's refugee physicists. Two factors came to bear: the first, a definite unknown, was how close the Germans were to having a bomb and the second, more apparent, was the many obstacles Allied scientists needed to overcome before they could succeed.
>
> The physicists on the mesa would have been shocked to learn the true state of affairs of nuclear research in 1944 Germany. It was nowhere close to developing a bomb.
>
> (219)

The above passages demonstrate the nationally organized performance of science and the scientist's role. Fermi, Oppenheimer, Feynman had personas of patriotic scientists so that scientific work and national identity, the pace of scientific work and national security, all become intertwined in the biography. In other words, nuclear science becomes a mode of fashioning a national identity, and the scientist was a part of this process. The nationalism and patriotism of the scientist was influenced by and contributed to this fashioning, the biographies imply.

In the immediate paragraph following the documentation of Compton's doubts about involving Fermi the immigrant in the national-security project of the atom bomb, the Fermi biography describes an anecdote:

> in Fermi's Columbia office, Compton asked him the question. Fermi promptly went to the blackboard and worked out, in Compton's own words, "simply and directly, the equations from which could be calculated the critical size of a chain-reacting sphere." Fermi then proceeded to estimate how much U-235 would be needed for a bomb.
>
> (172)

This description throws up two interesting features. The first and most obvious is that Fermi demonstrates his indispensability for the bombmaking project by writing out the equation for Compton. The second feature is the location of the account: Fermi-the-scientist is also an immigrant but his immigrant status, and his 'questionable' loyalty (it has been questioned by Compton, according to the biography) is subsumed under his scientific merit. That is, the truth or facticity of whatever Fermi wrote on the blackboard was interpreted by Compton – who had the power to accept or reject the immigrant's membership into the team – as demonstrating both: the scientific solution to the problem the bombmaking squad was experiencing and his commitment to the *American* project. In other words, Fermi's scientific acumen enables him to demonstrate his 'nationalist' loyalty to the American bomb project through his equations and calculations.

Elsewhere, Isidor Rabi, a Nobel Physicist, refused to move to Los Alamos and tried to argue the case with Oppenheimer: 'he did not, he told Oppenheimer, wish to make "the culmination of three centuries of physics" a weapon of mass destruction' (Bird and Sherwon 212). But Oppenheimer would underscore the *national* need for the bomb:

> "I think if I believed with you that this project was 'the culmination of three centuries of physics,'" he wrote Rabi, "I should take a different stand. To me it is primarily the development in time of war of a military weapon of some consequence. I do not think that the Nazis allow us the option of [not] carrying out that development."

Only one thing mattered now to Oppenheimer: building the weapon before the Nazis did.

(Bird and Sherwin 212)

The subject of nuclear patriotism would recur later too, years after the war and the Manhattan Project. At the 1954 hearing that would eventually strip him of access to the government projects and laboratories the focus of the hearing, by judges handpicked by Oppenheimer's bitter enemy, Lewis Strauss, the Chairman of the US Atomic Energy Commission, was Oppenheimer's so-called Communist past and connections and, by extension, his questionable nationalism, patriotism and commitment to the American hydrogen bomb project.

First, the final report of the hearing admits that 'there is no indication of disloyalty on the part of Dr Oppenheimer' and that it has seen 'eloquent and convincing testimony of Dr Oppenheimer's deep devotion to his country in recent years' (United States Atomic Energy Commission 1017). And yet it concluded in its 'Recommendation':

> The record shows that Dr. Oppenheimer has consistently placed himself outside the rules which govern others. He has falsified in matters wherein he was charged with grave responsibilities in the national interest. In his associations he has repeatedly exhibited a willful disregard of the normal and proper obligations of security.
>
> We find that Dr. Oppenheimer's continuing conduct and association have reflected a serious disregard for the requirements of the security system.
>
> We have found a susceptibility to influence which could have serious implications for the security interests of the country.
>
> We find his conduct in the hydrogen bomb program sufficiently disturbing as to raise a doubt as to whether his future participation, if characterized by the same attitudes in a Government program relating to the national defense, would be clearly consistent with the best interests of security.
>
> We have regretfully concluded that Dr. Oppenheimer has been less than candid in several instances in his testimony before this Board.
>
> (US Atomic Energy Commission 1019)

As Bird and Sherwin put it, 'the board deemed Oppenheimer a loyal citizen who was nevertheless a security risk' (540). And yet, Oppenheimer in his letter to K.D. Nichols dated 4 March 1954, ahead of his hearing at the Atomic Energy Commission, would construct a public selfhood: of a man committed to the American scientific scene and later the war effort. Describing his studies abroad, very early in his narrative, Oppenheimer writes:

In the spring of 1929, I returned to the United States. I was homesick for this country, and in fact I did not leave it again for 19 years. I had learned a great deal in my student days about the new physics; I wanted to pursue this myself, to explain it and to foster its cultivation.

(US Atomic Energy Commission 7)

To 'foster its cultivation' – by which Oppenheimer means cultivation of physics *in* the USA. He records how, when his other scientist friends 'went off to work on radar and other aspects of military research', he envied them (US Atomic Energy Commission 11). He adds: 'it was not until my first connection with the rudimentary atomic-energy enterprise that I began to see any way in which I could be of direct use [to the war effort]' (US Atomic Energy Commission 11). The persona of the patriotic atomic scientist, although called into question by the hearing, would endure until a decade later John F. Kennedy would honour Oppenheimer at the White House.

Charisma and Charismatic Authority

Hertha Pauli, journalist, and sister of Nobel Physicist Wolfgang Pauli (famous for his acerbic dismissal of particular theories with 'That isn't right. It isn't even *wrong*'), in a piece titled 'Nobel's Prizes and the Atom Bomb' in the December of 1945, cites Henry L. Stimson, US Secretary of War who said of J. Robert Oppenheimer:

[the success of the bomb project was] largely due to his genius and the inspiration and leadership he has given to his associates.

(unpaginated)

A little further Pauli writes: 'There was a universal feeling on the mesa: "Oppy [Oppenheimer] knows best"' (Pauli unpaginated). J. Robert Oppenheimer, the name most commonly associated with the making of the atom bomb – he was the brilliant leader of the project at Los Alamos – is introduced in this contemporary essay as 'still in his thirties, sensitive, somewhat dreamy, a man who collected Bach records and French Impressionists' (Pauli, unpaginated).

Pauli's portrait of the dreamy genius who alone, despite the numerous Nobel Laureates (Compton, Urey, Bohr, Laurence, Fermi, etc.) working on the Manhattan Project, knew exactly what to do in order to get the bomb ready would be an abiding theme in the portraits of the leader of the bombmakers.

Leslie Groves' initial impressions of Oppenheimer and then the change in them, as described by Ray Monk, also underscore great personal magnetism:

Oppenheimer was, compared to the people Groves had already met, a relatively junior member of the project. He was not, like Compton,

Fermi, Franck and Lawrence, a Nobel Prize winner; nor was he, like Szilard, Teller and Wigner, an originator of the atomic-bomb project. Moreover, he seemed, in his love of French poetry, his absorption in the literature of Hinduism and his resolutely theoretical approach to physics, the very personification of the remote academic whom Groves had come to despise.

And yet, on meeting the thirty-eight-year-old Oppenheimer, Groves was immediately won over, feeling that here, at last, was someone who could see and understand the real problems that the project faced.

(Monk 324–5)

Bird and Sherwin in their prologue use the term that would be associated with Oppenheimer throughout his life: 'a theoretical physicist who displayed the charismatic qualities of a great leader, an aesthete who cultivated ambiguities' (5). Over the years, this charisma was widely recognized by his students, peers and all who came in contact with him:

Oppenheimer had transformed himself through his work and his social life from an awkward scientific prodigy into a sophisticated and charismatic intellectual leader.

(Bird and Sherwin 179)

Bird and Sherwin underscore that Oppenheimer's charisma was due to the force of his personality, but also emanated from the quality of his work in physics. Immediately following the above description, they add:

It did not take long for those he worked with to be convinced that if the problems associated with building an atomic bomb were to be solved quickly, Oppie had to play an important role in the process.

(Bird and Sherwin 179)

Oppenheimer's biographers record those features that are readily identifiable with charismatic authority, the persona that would remain unchanged when anyone spoke of him:

Now I could see at firsthand the tremendous intellectual power of Oppenheimer who was the unquestioned leader of our group . . . The intellectual experience was unforgettable.

(Hans Bethe cited in Bird and Sherwin 179)

Jeremy Bernstein would also emphasize Oppenheimer's multiple skills – from physics to organizational problems – that shaped his charisma:

It was as if all his [Oppenheimer's] gifts had been hoarded for this occasion. The all but instantaneous ability to comprehend and synthesize scientific ideas, which had terrorized his students because they could not keep up with him, was now channeled to making the project work.

(unpaginated)

This was ability that led to personal authority and interventions, and both contribute to charisma, as Charles Thorpe and Steven Shapin put it in their study of charismatic authority and Los Alamos:

participants tell us that if we want to understand the social and technical order of one of the most important technoscientific sites modern times, we should get to grips with the role of embodied personal authority in general, and charisma in particular.

(549)

Expanding Thorpe and Shapin, one could argue, following commentators on charismatic leadership, that such a persona possesses four qualities:

(1) The leader must be viewed as superhuman or as a cultural ideal, (2) The leader's ideas are believed simply because *he* said them (contra-factual evidence is ignored), (3) Similarly, orders are obeyed solely because they emanate from the leader (no matter how incomprehensible or self- contradictory they may seem), and (4) Followers have an intense emotional attachment to the leader. Other characteristics of devotees include a belief that their leader is extremely energetic, utterly at ease in stressful situations, supremely learned, and totally confident that he is "destiny's child." The charismatic figure disdains bureaucracy and demonstrates impressive rhetorical skills, a knack for dramatic action, and may well be attributed with the ability to read minds, predict the future, heal or harm by his very presence, control the weather, and other magical feats. He also always has "magnetic eyes."

(Lindholm 41–2, emphasis in original)

All biographers record Oppenheimer's incredibly blue eyes, his rhetorical style and the speed of his intellect. Jeremy Bernstein tracing the reasons for Oppenheimer's stardom records the following: 'a combination of his association with this great new power-the atomic bomb-his striking looks, and his unusual use of language' (unpaginated). In fact, even as a teacher, the biographies record, his speed of talking and problem-solving was excessive:

Glenn Seaborg, later a chairman of the United States Atomic Energy Commission, complained of Professor Oppenheimer's "tendency to answer your question even before you had fully stated it." . . . Wendell Furry, who studied at Berkeley from 1932 to 1934, complained that

Oppenheimer expressed himself "somewhat obscurely and very quickly with flashes of insight which we couldn't follow."

(Bird and Sherwin 83–4)

Jeremy Bernstein, who as a student heard Oppenheimer speaking at Harvard, would attribute part of the latter's charisma to not just his command of the subject but also the diction and mode of communicating his ideas:

Nothing that has been written about his charisma as a public lecturer has been exaggerated. It was a mixture of phrasing that was both elegant and somewhat obscure. You were not quite sure what he meant, but you were sure that it was profound and that it was your fault that you didn't see why.

(Monk 641)

Those at Los Alamos were deeply attached to Oppenheimer and believed in his ability to solve the problems involved in the making of the bomb. This was the persona of the charismatic, authoritative scientist.

Speaking of Fermi, Segrè and Bettina Hoerlin write: 'his piercing eyes shone with the intensity of his intellect' (68). Even after he retired to St John in the Virgin Islands, Oppenheimer's charisma seemed unchanged: 'Despite his cultivated aura of otherworldliness, he fit comfortably into their island world' (Bird and Sherwin 573).[9] Ray Monk's biography also quotes Isador Rabi's comments on Oppenheimer's otherworldly air:

"In Oppenheimer," Rabi remarked, "the element of earthiness was feeble.

"Yet it was essentially this spiritual quality, this refinement as expressed in speech and manner, that was the basis of his charisma. He never expressed himself completely. He always left a feeling that there were depths of sensibility and insight not yet revealed. These may be the qualities of the born leader who seems to have reserves of uncommitted strength."

(Monk 669–70)

Years later, Oppenheimer would say exactly the same of Albert Einstein: 'he was almost wholly without sophistication and wholly without worldliness' (38).

Oppenheimer signalled his charisma, according to the biographies, in particular ways. I use the concept of 'signaling charisma' from Nicolas Bastardoz (2021):

An effective way to achieve her aim [of conveying certain attributes about herself] may be to communicate the values she stands for and indicate she has understood the group's concerns and shares her followers' emotions, all this wrapped up in symbolic rhetoric. The charisma signal arises out of such communication.

(313)

He adds that 'the signal in itself has *per se* no (or very little) intrinsic value; its value lies in the credible information about the underlying attribute' (314) and that 'Charisma is . . . both a signal of a leader's ability (i.e., high intelligence) and intentions (i.e., the focus on value and group emotions)' (315).

Oppenheimer begins signalling his charisma at Los Alamos through, first, his capacity for organization (itself a surprise to many, since he was not known for organizational skills). Bird and Sherwin record:

> However disconnected from his responsibilities Oppenheimer may have seemed before he moved to Los Alamos, he quickly demonstrated his capacity for change. [Robert] Wilson was surprised after several months at Los Alamos to see his boss metamorphose into a charismatic and efficient administrator.

More vexing was the issue of military control over the scientists and the laboratories. As Leslie Groves made clear to everyone, including Oppenheimer, the military was the supremo but the scientists 'flatly opposed the notion of having to work under military discipline' to the Scientific Director, Oppenheimer. After several parleys and negotiations, Oppenheimer managed a key compromise.

> During the lab's experimental work, the scientists would remain civilians, but when the time came to test the weapon, everyone would don a uniform. Los Alamos would be fenced and designated an Army post – but within the "Technical Area" of the lab itself, the scientists would report to Oppenheimer as "Scientific Director." The Army would control access to the community, but it would not control the exchange of information among the scientists; that was Oppenheimer's responsibility.
>
> (Bird and Sherwin 211)

This administrative move was a coup and Hans Bethe, head of the theoretical division at Los Alamos, said to Oppenheimer: 'I think that you have now earned a degree in High Diplomacy' (Bird and Sherwin). These successful negotiations in the biographies' construction of the persona of a charismatic Oppenheimer also revealed to his team his values and underlying attributes.

By ensuring that the laboratories and the scientists were *not* under the military's jurisdiction and control, Oppenheimer signalled the primacy of science even in the midst of war effort. His commitment to the science of the bomb was made clear in the move to secure administrative, communicative and spatial (laboratory) freedom for his peers and scientists – and this was precisely what Bethe was congratu-lating Oppenheimer about in the above passage.[10]

In the Bird-Sherwin biography, the authors focus on the organizational issues that hampered the scientists and point to the values with which Oppenheimer operated and managed to communicate to his group. These issues, again, stemmed from the situation where the army was in control of the Project but so were the scientists. The insistence on secrecy, com-partmentalization of information, censorship irked the scientists no end, as all the biographies point out. When reporting the concerns around the implosion mechanism, Bird and Sherwin document the clash between the ordnance commander, Captain Parsons, and the scientists, George Kistiakowsky and Seth Neddermeyer. The account of the clash comes immediately after the statement of the scientific-technological problem: having noted that America needed to test a low-efficiency weapon before going in for the larger more effective one, Bird and Sherwin write of the stalled work: '[Seth] Neddermeyer and his men in the Ordnance Division were making very little progress on the implosion design' (279). Bird and Sherwin then go on to describe the incident as follows:

> Parsons chafed at what he considered a loss of control over his Ordnance Division, and in September he sent Oppie a memorandum proposing to give himself broad decision-making powers over all aspects of the implosion bomb project. Oppenheimer gently but firmly refused: "The kind of authority which you appear to request from me is something I cannot delegate to you because I do not possess it. I do not, in fact, whatever protocol may suggest, have the authority to make decisions which are not understood and approved by the qualified scientists of the laboratory who must execute them." As a military man, Navy Captain Parsons wanted the authority in order to short-circuit the debates among his scientists. "You have pointed out," Oppenheimer wrote him, "that you are afraid your position in the laboratory might make it necessary for you to engage in prolonged argument and discussion in order to obtain agreement upon which the progress of the work would depend. Nothing that I can put in writing can eliminate this necessity." The scientists had to be free to argue – and Oppenheimer would arbitrate disputes only for the purpose of reaching some kind of collegial consensus. "I am not arguing that the laboratory should be so constituted," he told Parsons. "It is in fact so constituted."
>
> (280–1)

From the above passage, several aspects of the Los Alamos operations emerge, and Oppenhemier's role in them. The emphasis on control – military or scientific? – is a matter of considerable dispute, evidently. Then, Oppenheimer makes it clear that the powers of decision-making about the implosion bomb were vested with the scientists and therefore *their* statements, claims and decisions were to be taken to be of value rather than what the military may rule. Third, the laboratory was a scientific space and as such beyond the jurisdiction of the military.

With successful negotiations such as the above, in keeping the military and the science separate, Oppenheimer was able to secure the loyalty and the command of his group at Los Alamos. Much of this attachment stemmed, as Max Weber had argued in his theory of charismatic authority, from the high degree of personalness with which Oppenheimer operated.

Robert Wilson said:

> When I was with him, I was a larger person . . . I became very much of an Oppenheimer person and just idolized him . . . I changed around completely.
>
> (Bird and Sherwin 205)

Even the administrative staff were mesmerized by Oppenheimer. One of them, Dorothy McKibbin, responded to his charisma:

> McKibbin was immediately smitten by Oppenheimer's easy grace and charming manners. "I knew that anything he was connected with would be alive," she recalled, "and I made my decision. I thought to be associated with that person, whoever he was, would be simply great! I never met a person with a magnetism that hit you so fast and so completely as his did. I didn't know what he did. I thought maybe if he were digging trenches to put in a new road, I would love to do that. . . . I just wanted to be allied and have something to do with a person of such vitality and radiant force. That was for me.
>
> (Bird and Sherwin 63)

His informality as Director endeared him to most people. Bird and Sherwin describe how Oppenheimer would arrive at his office:

> As Oppie walked to the "T," [Technical Area] his colleagues often fell in behind him and listened quietly as he softly murmured his thoughts of the morning. "There goes the mother hen and all the little chickens," observed one Los Alamos resident. "His porkpie hat, his pipe, and something about his eyes gave him a certain aura," recalled a twenty-three-year-old WAC who worked the telephone switchboard. "He never needed to show off or shout. . . . He could have demanded Priority One with his telephone calls but never did. He never really needed to be as kind as he was." His informality contrasted sharply with the manner of General Groves, who "demanded attention, demanded respect." Oppie, on the other hand, got attention and respect naturally.
>
> (216)

Charismatic authority, as it is detailed in the biographies especially of Oppenheimer, was central to the persona of the celebrity scientist.

The Moral Scientist

David Hecht, commenting on the Oppenheimer 'defence' presented at the 1954 hearing, argues that

> Oppenheimer's autobiographical self-defense demonstrates the central role of personal narrative in establishing cultural authority ... it also functions as an argument that personal narrative was both relevant and essential to adjudicating the issues of credibility and loyalty raised by the charges against him.
>
> (169)

Hecht is commenting on the cultural authority and the public role of the atomic scientist. This authority and role, in many cases as we shall see, emerged from the moral stance and activism against nuclear armaments and technology. The biographies, after developing the epistemic virtue in the persona of the atomic scientist, invariably pay some attention to the political virtues, aimed, as Herman Paul would argue, at the acquiring of moral and political goods such as freedom and security. It is to this aspect of the biographies that I now turn.

The Franck Report (1945), Oppenheimer's speech to the Los Alamos scientists (1945), the Bohr letter to the United Nations (1950) and other key documents cited and examined in the biographies enable the biographers to paint the persona of the tormented Oppenheimer (he had endorsed the use of the bomb against human populations) and those morally opposed to the bomb.

Ray Monk writes about the cultural work of the Franck Report:

> The report was an extraordinarily farsighted and persuasive document that demanded to be taken seriously, not only because of the intrinsic merits of its arguments, but also because it was written by scientists who had been central to the development of the atomic bomb from the very beginning and who understood, as well as anyone, its destructive power.
>
> (434)

Silvan Schweber, in his comparative biographies of Einstein and Oppenheimer, writes:

> The creation of nuclear weapons had made warfare "inescapably a civilian as well as a military affair" and had made obsolete the separation and isolation of the world of science and of the intellect from that of politics and practical affairs. Oppenheimer came to symbolize

these changes, and was acknowledged as having the moral courage to act accordingly.

(4)

Both Monk and Schweber point to the moral debates around the bomb. It is in the context of the scientists' campaigns and public pronouncements before and after the bombing of Hiroshima–Nagasaki that the moralization of nuclear energy and nuclear technology in the scientific biographies needs to be read.

Scientists like Arthur Compton sought the moral high ground for the USA in curbing atomic armaments. Compton in 1946 wrote to Niels Bohr:

> What the world needs most just now is for the United States to take decisive moral leadership. For generations the nations have looked to the United States as a land of hope, of freedom and good will. That view is not gone, but it has been shaken . . . A decisive moral act by the United States now would answer these questionings. New hope would stir.
>
> (Cited in Schweber 73)

Compton was one of many who spoke out about the terror of nuclearization. Writes Gleick:

> As physicists began to speak out about world government and the international control of nuclear arms, so an army of clerics, foundation heads, and congressmen now made the mission and the morality of science a part of their lecture-circuit repertoire.
>
> (unpaginated)

In the case of Oppenheimer, part of the persona of a regretful scientist stems not from the moral position about controlling atomic armaments that he espoused later alone but from the deep *ambivalence* that marks the man, an ambivalence that biographers struggle to grapple with. Even before the bomb, Oppenheimer had expressed concerns:

> The goal of atomic energy, he [Oppenheimer] said, should be the enlargement of human welfare, and America's moral position would be greatly strengthened if information were offered before the bomb were used.
>
> (Cited in Lanouette 274)

Pondering over the question as to why, when his views had diverged from that of the US government in the 1950–52 period, he continued to work as a consultant to it, Ray Monk offers this:

> Oppenheimer continued to act as a consultant to government projects, thereby exposing himself to all sorts of exhausting conflicts

and crushing unpleasantness, precisely because of his love of, and loyalty to, his country. He did it for the same reason that he underwent the extreme rigors of leading Los Alamos: because he felt that it was, using the word that underpins the morality of the Bhagavad Gita, his duty to do it.

(579)

Monk iconizes, I suggest, a morally divided Oppenheimer.[11] Oppenheimer, in Hecht's astute reading too, 'makes a moral argument' (175–6).

Oppenheimer here is not an icon that inspires awe for his brilliance alone but one who elicits awe for being courageous enough to tackle the government in the face of disregard, suspicion and anger (Monk records just *before* the above passage that 'Oppenheimer was disliked, or at the very least held in suspicion by nearly every person in high office in Washington', 579).

Even before the Trinity test and the Hiroshima–Nagasaki explosions, some of the scientists involved had been wary of what they were inventing. For instance, the Szilard biography records his uneasiness at the successful test in Chicago of the uranium pile's criticality:

> To Szilard, who had feared this moment for more than nine years – and had hoped all that time that it would not occur – his colleagues' exuberance must have seemed unnerving. He had first conceived the chain reaction in 1933 in a burst of defiant creativity. He had restrained his ego and energies for more than two years while collaborating with Fermi to design and build the first reactor. And now Szilard had witnessed his vision come true. As he had feared from the start, the world would never be the same again.
>
> After the experiment and the silent sipping of wine, as their colleagues filed from the squash court into the cold evening light, Szilard and Fermi found themselves standing alone. "I shook hands with Fermi," Szilard remembered, "and I said I thought this day would go down as a black day in the history of mankind."
>
> (Lanouette 252)

The success of their efforts and the proof of their scientific theories is bitter-sweet, and the scientist demonstrates, as the biography underlines, a moral dilemma, even a crisis: what exactly have they invented and achieved? This is a moral question about their years of science, but also about the future they have just inaugurated (as Szilard's prophetic words above indicate).

Herman Paul has argued in the case of historian-scholars that

> the interesting question is no longer *whether* historians maintain a moral relation to the past, which they surely do, but *which* moral goods (equality, freedom, respect, and so on) they pursue in the

course of their studies and how those goods relate to other (epistemic, aesthetic, political) goods they seek to acquire.

(363, emphasis in original)

The goods sought by many of the celebrity scientists and former bombmakers, the biographers note, was peace and international collaboration aiming at more 'openness' (Niels Bohr's pet theme) and control over nuclear technology and weapons.

Alice Kemball Smith, writing about the Franck Report, notes the 'accuracy with which the framers of the Franck Report forecast the course of postwar armaments race' (45). She admits that 'this is not to say that scientists are possessed of universal political wisdom' (45) but that 'Franck and his colleagues . . . concluded that the cause of peace would in the long run best be served by demonstration of the bombing before dropping it as a last resort upon Japan' (45). The point however is not the persona of the scientist as a political visionary but one whose intellect – which enabled his work on the bomb – is now directed at foreseeing the havoc likely to emerge from the bomb's use: a foresight that is at once scientific (being able to understand the exact nature of atomic power) and moral.

In his account of Harold Urey's life, Matthew Shindell, drawing the persona of the concerned scientist, tells us that Urey's concerns over militarization was always a part of his speeches and talks:

by the late 1930s Urey was convinced that two fates were possible. One was the further militarization of science. Here – foreshadowing his postwar concerns over atomic weapons – he imagined that science's support of military activity could bring about "the complete destruction of our civilization.

(unpaginated)

And:

For the first decade after the end of the war, Urey attempted to balance his politics and aversion to weapons work with his scientific research and its reliance on government and military funding.

(Unpaginated)

Later, Shindell would dwell approvingly on the moral stance of Urey and other atomic scientists:

For Urey, and for many of the scientists involved, atomic weapons demanded that the world be reorganized [because] No one who understands atomic war wants anything but peace.

(Unpaginated)

In the process of developing this persona of his biographical subject, Shindell does indicate a degree of naivete in him. Citing Urey's own writings, Shindell says:

> In a moment of naivete, Urey imagined a meeting of Russian and American leaders, along with their scientific advisers, in which the scientists acted as mediators between the two groups: "Scientists will have no trouble understanding one another. When they meet I think their recommendations will be almost unanimous."
>
> (Unpaginated)[12]

Lanouette cites Pierre Auger, the nuclear physicist, who says of Szilard in 1945:

> Szilard was politically preoccupied at the time about the future of humanity and the organization of a proper type of republic.
>
> (254)

Szilard's memo to Franklin Roosevelt, 'Atomic Bombs and the Postwar Position of the United States in the World' would be described as 'a visionary document' by his biographer (267).

The atomic scientist, as can be seen from these biographies, was also in many cases, a moral scientist, or more accurately, a morally conflicted scientist.

The authority and aura of the atomic scientist, made possible by the various layers of persona they inhabit, and made possible by their circle and networks of peers and institutions, may be said to constitute both scientific and cultural work. The biographies then produce a persona beyond the traditional one of the 'maker-of-the-bomb'.

Notes

1 Years later, in its inaugural issue of May 1948, *Physics Today*, from the American Institute of Physics, had as its cover a hat sitting on a piece of machinery, but no names were mentioned. The machine was a cyclotron, and the hat was a pork pie hat – made famous by Oppenheimer. The cover made it clear that the name and identity was superfluous: this was the hat worn by the most famous physicist in the USA. David Hecht refers to the making of the 'public Oppenheimers' in his essay of 2008.

2 I write this in the month that Christopher Nolan announced the launch of *Oppenheimer*, starring Cillian Murphy in the title role, and based on the Bird-Sherwin biography.

3 These skills were, in cases such as Oppenheimer's, expansive beyond their disciplines. For instance, Bird and Sherwin record the feelings of Karl T. Compton (the future President of MIT) when he encountered Oppenheimer:

Compton . . . felt intimidated by Oppenheimer's extraordinary versatility. He could hold his own with the young man when the topic was science, but found himself at a loss when Robert began talking about literature, philosophy or even politics. (59)

4 For a history of 'nuclear secrecy', see Wellerstein (2021). Robert Jay Lifton and Greg Mitchell spend a considerable part of their *Hiroshima in America* examining the 'official narrative' after the bombings, particularly the silence of the scientists who built the bomb. They also note that despite the official pressure to not express their views on the political aspects of the bombing, scientists like Leo Szilard did try to have a public discussion around the bombings (see Chapters 5 and 6).

5 That Oppenheimer himself was not a Nobel Laureate did count against his appointment as the Director. Leslie Groves writes:

Oppenheimer had two major disadvantages – he had had almost no administrative experience of any kind, and he was not a Nobel Prize winner. Because of the latter lack, he did not then have the prestige among his fellow scientists that I would have liked the project leader to possess. The heads of our three major laboratories – Lawrence at Berkeley, Urey at Columbia, and Compton at Chicago – were all Nobel Prize winners, and Compton had several Nobel Prize winners working under him. There was a strong feeling among most of the scientific people with whom I discussed this matter that the head of Project Y should also be one.

(62)

6 The commissioned portrait of the scientist, as Patricia Fara has argued,

reflect and create exemplary ideals, simultaneously depicting, establishing, consolidating and advertising ideological constructs such as national character, appropriate gender behaviour and class structures . . . Portraits show an individual, yet simultaneously reveal the face of science itself.

(79)

It should be possible to see the photographs, consciously staged and shot, as offering resonances with the portrait for the same reason, although this is not the subject of the current chapter.

7 Years later, of course, when Oppenheimer is arraigned, this role of secret-unraveller will be again underscored. *The New York Times*, for instance, would note: 'Dr Oppenheimer now carries around in his head as much top secret information as any man alive' (cited in Hecht 956).

8 At the hearing, it would be this very patriotism and national/ist priorities and attitude as a scientist that would be called into question in the post-war period by the Board:

We find his conduct in the hydrogen bomb program sufficiently disturbing as to raise a doubt as to whether his future participation, if characterized by the same attitudes in a Government program relating to the national defense, would be clearly consistent with the best interests of security.

(Bird and Sherwin 541)

9 The otherworldliness of the scientist is most explicitly seen as a theme in Eve Curie's *Madame Curie* (1937), where she writes not only about Curie's frugal lifestyle but also the refusal to patent radium or make money from her discovery.

10   Reading the Oppenheimer hearing, Charles Thorpe comments that the entire exercise was about the legitimacy of science. Thorpe writes:

> The hearing was a contest for legitimacy and for the right to define the cultural and political role of science. It was an event which crystallized tensions between competing understandings of the legitimate place of scientists and scientific expertise in the operations of the state and in civil society.

(528)

11   Elsewhere, David Hecht frames Oppenheimer's growth into a scientific icon:

> Iconic figures do not emerge randomly; they reflect the culture that shapes them. Depictions of Oppenheimer as heroic were no exception. He proved to be an apt scientific icon for many Americans made anxious by unsettling developments in the atomic age. I argue that non-scientific attributes were the basis of this appeal; Oppenheimer's evident humanism was a prominent example of a larger phenomenon in which his admirers elevated him to heroic status by contextualizing his technical achievements amid an array of reassuring personal traits. Portrayals might emphasize a variety of roles for Oppenheimer: as humanist, moralist, patriot, intellectual, adventurer, or activism.

(944)

Like Hecht, I use the term 'icon' by shifting it away from its religious connotations and more on the lines of the 'secular icon' as defined by Vicki Goldberg, Cornelia Brink, among others. Goldberg writes:

> I take secular icons to be representations that inspire some degree of awe – perhaps mixed with dread, compassion, or aspiration – and that stand for an epoch or a system of beliefs . . . I think of as icons almost instantly acquired symbolic overtones and larger frames of reference that endowed them with national or even worldwide significance. They concentrate the hopes and fears of millions and provide an instant and effortless connection to some deeply meaningful moment in history. They seem to summarize such complex phenomena as the powers of the human spirit or of universal destruction.

(Goldberg 135)

12   Shindell notes that when he needed funding for his instruments and equipment for researching geology,

> 'Urey put his aversion to military contracts aside and made his first contract proposal to the O[ffice] of N[aval] R[esearch], asking for about $105,000 for an "investigation of natural abundances of stable isotopes with the primary objective of measuring paleo-temperatures"'.

(Unpaginated)

Shindell adds: '1953/54, Urey received $55,956 from the US Atomic Energy Commission and $21,400 from the National Science Foundation' (Unpaginated).

4 Irradiated Aesthetics
The Atomic Sublime

In *Hiroshima Diary*, the physician Michihiko Hachiya ponders:

> What kind of a bomb was it that had destroyed Hiroshima? What
> had my visitors told me earlier? . . .
> Otherwise why did the alarm stop? Why was there no further alarm
> during the five or six minutes before the explosion occurred? . . .
> Whatever it was, it was beyond my comprehension. Damage of
> this order could have no explanation! All we had were stories no
> more substantial than the clouds from which we had reached to
> snatch them.
> One thing was certain – Hiroshima was destroyed; and with it the
> army that had been quartered in Hiroshima.
>
> (25)

Hachiya documents, alongside the massive destruction of his city,
an inability to comprehend the nature of what happened. In a similar
fashion, John Hersey, in *Hiroshima*, reports the responses to the explosion by a resident:

> From the mound, Mr. Tanimoto saw an astonishing panorama. Not
> just a patch of Koi, as he had expected, but as much of Hiroshima as
> he could see through the clouded air was giving off a thick, dreadful
> miasma. Clumps of smoke, near and far, had begun to push up
> through the general dust. He wondered how such extensive damage
> could have been dealt out of a silent sky . . .
>
> (45)[1]

Hachiya also uses the term 'panorama' to describe the sights of devastation (193).
  The expanse and scale of destruction can be perceived but not understood, and so 'wonder' and 'incomprehension' are emphasized in eyewitness accounts of the events, such as the above, at Hiroshima and
Nagasaki. The atomic bomb is the unthinkable that has happened, and
hence the stark incomprehension. Moreover, atomic effects are never

DOI: 10.4324/9781003254294-4

localized nor restricted in terms of time. Svetlana Alexievich writes in *Chernobyl Prayer*:

> We're still using the old concepts of 'near and far', 'them and us'. But what do 'near' and 'far' actually mean after Chernobyl, when, by day four, the fallout clouds were drifting above Africa and China? The earth suddenly became so small, no longer the land of Columbus's age. That world was infinite. Now we have a different sense of space.
>
> (24)

Alexievich is signalling the sheer scale of atomic effects.

To underscore the all-pervasive radiation in their lives, in the series *Chernobyl*, as the dying Vassily and his wife embrace in the ward of Hospital 6, the screen is suffused with a glow (episode 3). Adam Higginbotham in *Midnight in Chernobyl* cites Senior Unit Engineer Boris Stolyarchuk:

> Dawn had broken. The light was crisp and clear. What Stolyarchuk saw did not frighten him, but he was struck by one thought:
>   I'm so young, and it's all over.
>   Reactor Number Four was gone. In its place was a simmering volcano of uranium fuel and graphite – a radioactive blaze that would prove all but impossible to extinguish.
>
> (110)

The sight is sublime, the viewer senses something impossible, uncontrollable, that has occurred.

Such instances of epistemological, cognitive and affective failures in the face of witnessing and experiencing the horror of a nuclear devastation in the prose, poetry and visual texts of the Hiroshima–Nagasaki bombings, the nuclear tests in Australia, Marshall Islands and Nevada, and the disasters at Fukushima and Chernobyl call for their own aesthetic strategies.

The broadest possible mode of/in irradiated aesthetics would be that of the sublime. Theorized as the 'nuclear sublime' by Frances Ferguson (1984), Joseph Masco (2004) and Michael Shapiro (2018) or the 'atomic sublime' by Peter Hales (1991), the sublime in nuclear accounts and literature, including *hibakusha* poetry (Miller and Atherton 2017), however, is not just of the apocalyptic (Gunn and Beard 2000) variety. Hales, for example, argues that the atomic sublime – which emphasized the awe at the spectacular nature of atomic explosions – shades into the 'atomic *gothic*' wherein 'the witness was victim and not spectator [and] the dominant psychological state was not awe and pleasure, but pain' (24, emphasis in original). Hales recognizes that to see, visualize and speak of the atomic explosions as mere spectacle is to centre an 'abstract visuality' – Shapiro's term (85) – rather than the human (and ecological, although Hales does not go so far) costs of the explosion. The 'celebratory comprehension' (Shapiro 85) of such an abstract visuality is modified through a shift in emphasis in certain texts, notably Japanese, who encountered the sublimity of destruction *on the ground*. That is, it is not a sublimity of safe, distant viewing of the disaster alone: it is *also* a sublime of highly personalized, subjectivity-destroying encounter with the unthinkable, and the lingering, iterative effects of that encounter.

The sublime is the 'unthinkable' (Ferguson 5) and 'the nuclear sublime, then, operates much like most other versions of the sublime, in that it imagines freedom to be threatened by a power that is consistently mislocated' (Ferguson 9). Reading postmodern American poetry, Rob Wilson identifies a nuclear sublime that 'whether sensed as a complex presence or unarticulated absence, comprises the terror of a technological determination of the Cold War period reducing the subject to languagelessness, almost bodily sublation' (409). Joseph Masco aligns the sublime appeal with an excess of danger: 'if the conceptual force of the sublime is directly proportional to the danger involved in the experiential event . . . then nuclear weapons offer access to a uniquely powerful manifestation of sublimity' (3). Rebecca Solnit writes of the contemporary sublime:

> the unnatural disasters of the present offer no such containment within the bounds of the natural – the oil fields afire in Kuwait, the mushroom clouds above Yucca Flat, the blood-red sunsets of Los Angeles – though they still compel attention . . . this new, horrible sublime . . .
>
> (unpaginated)

But the nuclear sublime also represents a massive shift. Where once the sublime was about the power of nature to defeat language and human imagination, with the Holocaust, we see 'sublime effects among our own responses to this demonstrated human potential for systematic and unbounded violence . . . human-inflicted disaster will remain more threatening, more sublime, than any natural disaster' (Gray 1). In resonance with Gray, Henry J.M. Day writes:

> the obliteration of Hiroshima by the first atomic bomb undermined faith in the redemptive potential of the divine, the natural world or one's own humanity at the same time terrifyingly reinstantiated the sublime as a function of traumatic world historical events.
>
> (Cited in Miller and Atherton 3)

By proposing traumatic 'world historical events' as instantiating the new sublime, these critics also reconstitute the sublime as transhistorical and trans-local, just as the 'thick, dreadful miasma' that the Hiroshima resident describes is resonant with the 'the fallout clouds . . . drifting above

Africa and China' in Alexievich's account of Chernobyl. They instantiate the sublime in rhetoric from different nuclear events and topoi. They therefore provide the evidence for a transhistorical sublime aesthetic.

Patrick Wright persuasively argues for the transhistorical nature of the sublime in the modern era, as a 'rhetorical style' (88), and 'a modern restructuring of the sacred' (92. Also see Blernoff 2002for the sublime as a transhistorical category). Linking trauma with aesthetic of the sublime, Dylan Sawyer writes:

> Rather than evoke the balanced pleasure/pain synthesis of the traditional sublime, I believe that the magnitude of the traumatic event – the event *having been recognised* to have instigated trauma – weights its victims with resentment and despair at the impossibility of providing any adequate response . . . The traumatic event is to be understood then as a persistent proximal presence unable to provide the 'safe-distance' necessary to evoke a sense of the 'traditional' sublime, overwhelming the perspective of its victim to ensure that while he or she cannot escape trauma's shadow, neither are they able to sharpen it into any clear definition. Nevertheless, although the victim can ultimately never know the event itself (since it is inherently unknowable), he or she can know, or at least can still *engage* in attempting to know, the traumatic event since it is the subject's own enquiry (impaired though it is) that constantly brings it into being.
>
> (171, emphasis in original)

For Sawyer, the *traumatic sublime* is 'a sensation constituted more by the *despair* felt at the mind's failure to appropriate the event's full magnitude combined with an acknowledgement of its unceasing attempt to recreate and represent the traumatic event's occurrence' (173). When trauma is repeatedly invoked in the form of memories or testimony, it approximates to the structure of the sublime:

> The traumatic sublime is therefore to be understood not simply as a response to the event but more as the recurring event of such a response, attesting to *inability itself* appearing partial and the cause for further instances of trauma (and perhaps testimony). The traumatic sublime is a sensation unable ever to be sated or savoured and one that ultimately finds more impetus from impossibility than possibility itself.
>
> (173, emphasis in original)

Sawyer's argument enables me to read the texts of nuclear disaster and effects because he points to continuity rather than a limited 'event'. Given the fact of the *continuity* of radiation effects upon human bodies, the planet's non-living materials, the non-human forms of life, the iterations of sickness and contamination that that – which the Rand Corporation report,

Worldwide Effects of Atomic Weapons: Project Sunshine (1956) and later Chernobyl's toxic fumes spreading across Europe also demonstrated – the nuclear sublime as irradiated aesthetics captures a state of *permanent crisis*, whose scale of damage is at once visible/visual (marked *upon* bodies and objects) and invisible (in terms of bodies poisoned *within*). The nuclear sublime is more than the aesthetics of an event or its descriptions in visual and verbal texts: it is a horror-aesthetics of a continuing crisis and permanent damage across the world. How exactly such a transhistorical and trans-local aesthetics operates is the subject of this chapter.

Examining a range of verbal and visual texts about Hiroshima–Nagasaki, Fukushima and Chernobyl, this chapter unpacks the nuclear sublime. It does not privilege the sublime's distanced view – which is the case with the sublime employed when people view the terrifying spectacle from afar and from safety, and are overawed by the sight – and tempers such a view with the close, and haunting, encounters with mass death, the spectacle of large-scale injuries and devastation of habitations and neighbourhoods.[2]

The Atomic Sublime

A *prefiguration* of the atomic and catastrophic sublime is visible in official accounts and representations of the power of the bomb in the lead-up to the Hiroshima–Nagasaki explosions: in the Trinity tests. This aesthetic of documentary realism focuses attention on the *making* of the bomb in a non-sentimental manner even as it references the destructive potential of what the bomb *may* unleash. Resonant with the 'fabulously textual' nature of nuclear war that Jacques Derrida identifies in his 1983 essay on 'nuclear criticism', these images of the military-nuclear complex – Chicago, Oakridge, Los Alamos – are an 'abstract visuality' (Shapiro 85) or 'radioactive visuality' (Ganguly 438) which can only be read in conjunction with the later images of the destruction of the cities of Hiroshima–Nagasaki.

The atomic sublime here is, in fact, an atomic-industrial sublime.

An Atomic-Industrial Sublime

Richard Rhodes' *The Making of the Atomic Bomb* carries photographs of the Curies, Rutherford, Szilard, Hahn and Meitner and other scientists, several of them photographed at work in their laboratories. He also gives us images of the laboratories where the atom and eventually nuclear power was studied and harnessed: the Cavendish (Cambridge), the then Kaiser Wilhelm Institute (Munich), the Institute for Theoretical Physics (Copenhagen) and others. In order for us to understand the 'why' of the atomic bomb research in Euro-America, he places images of Hitler, Pearl Harbour and Coventry Cathedral destroyed by German bombs. It is only *after* these images of the destruction of English and American 'property' that Rhodes places the photographs of the sites of atomic power and technology: Chicago, Oak Ridge, Hanford, Santa Fe, implying a cause–effect sequence. Later photographs include the Trinity test preparations and the 16 July 1945 explosion which finally revealed what the atom contained. Rhodes does include the Hiroshima–Nagasaki explosions and images of the widescale destruction and later tests in the Marshall Islands – by which time, of course, the power of nuclear energy had been made visible already.

Images of laboratories and factories in histories of the bomb, such as Rhodes' or in biographies of the nuclear scientists, are prefigurations of the atomic sublime because, while on one hand they show innocuous buildings and/or spaces where pure science is pursued, on the other, they constantly highlight what is at stake in the processes undertaken within these spaces. These histories of nuclear power and technology with their photographic-textual framing signal the sublime lying in wait inside the atom. Edmund Lee and Shirley Ho examining the modes of photographic-textual framing, argue that

> the use of photographs and texts must have been selected and organized in such a way as to make salient an aspect of a perceived reality . . . the context of the photographic–textual frame should be explained by accompanying texts . . . the presence and length of the textual element must not significantly shift the focus away from the photograph.
>
> (Lee and Ho 953)

Rhodes' captions frame the photographs of Hanford, Chicago and other laboratories:

> Chicago Pile Number One, the first man-made nuclear reactor, under construction at the University of Chicago, November 1942.
>
> (Rhodes, image 53)

> U.S. plutonium-production complex on the Columbia River at Hanford, Washington. Twelve-hundred-ton graphite reactors drilled with 2,004 channels held uranium slugs; neutrons from fission transmuted 250 parts per million of U238 to plutonium. D pile in foreground between water tanks.
>
> (Rhodes, image 58)

The factories are themselves an instance of the industrial sublime, in terms of sheer scale:

> K-25 gaseous-diffusion plant, Oak Ridge, Tennessee. Built to monumental scale, the structure is half a mile long with 42.6 acres under roof.
>
> (Rhodes, image 55)

Leslie Groves in *Now It Can Be Told* includes images and descriptions of the sites where the bomb was manufactured. His emphasis, like Rhodes', is on the scale of operations.

These requirements [for the Hanford site] were:

An estimated 25,000 gallons per minute of water would be needed, assuming recirculation of cooling water.

An estimated 100,000 kw of electricity would be required, with favorable load and power factors.

The hazardous manufacturing area should be a rectangle of approximately 12 miles by 16 miles.

The laboratory should be situated at least 8 miles away from the nearest pile or separation plant.

The employees' village should be no less than 10 miles upwind from the nearest pile or separation plant.

No town of as many as 1,000 inhabitants should be closer than 20 miles to the nearest pile or separation plant.

(Groves 71)

The images of the Hanford sites are organized into two: before the factory is built, and after. The caption says:

Typical Hanford scene (top) before construction; above, a completed reactor at Hanford in operation; left, construction workers at Hanford; and opposite page, top, the Man ford construction camp, which holds 45,000 workers.

(Groves, unpaginated)

There is a panoramic view of the massive Oak Ridge unit, some photographs of the workers and armed guards at the various plants, and of course of the scientists behind the operation.

Taken together, the photographs and the photographic-textual framing communicate certain key aspects of the reality: that the harnessing of nuclear energy was in and of itself a massive project involving multiple locations and thousands of people (there are mentions of the money spent as well, in Groves and Rhodes); the science and technology involved were sublime in terms of sheer quantities and power (note Rhodes' insistence on numbers). Images and descriptions of the New Mexico or Hanford (Benton) landscape, desert, the Columbia river and the mesa are juxtaposed with the state-of-the-art technological units set up therein. All this, I suggest, prefigures the atomic sublime as an atomic-industrial sublime, because the images and texts capture the vastness of the project and the monumental power the projects were trying to harness. Thus, the accounts and images bring together vast landscapes and technology, as Michael Shapiro proposes in his account of America's industrial sublime (104).[3]

The atomic-industrial sublime is the nuclear-military complex. The complex encapsulates the spaces, the routines and the processes that made the power available, and in many cases, the end-product (the bomb) as well. The atomic-industrial sublime is one of dispossession – the native population in Nevada, New Mexico either ignored as 'down winders' or moved off the land – and land management. If Laurence spoke of the elemental powers of nature being harnessed through atomic plants and the bomb, the industrialization of nature depended on this very nature for its functions: the presence of Columbia's water, for example, was crucial for Hanford. As Groves puts it:

> The plateau on which the plant would be built was only a few miles from the Columbia River, which had a superabundance of very pure and quite cold water. The site was well isolated from near-by communities, the largest of which was the town of Pasco, and if an unforeseen disaster should occur, we would be able to evacuate the inhabitants by truck.
>
> (75)

And about New Mexico:

> From the standpoint of security, Los Alamos was quite satisfactory. It was far removed from any large center of population, and was reasonably inaccessible from the outside. There were only a few roads and canyons by which it could be approached. Also, the geographically enforced isolation of the people working there lessened the ever-present danger of their inadvertently diffusing secret information among social or professional friends outside.
>
> The only major problem left was whether the school's owners would object to its being taken over. It was a private school with students from all over the country and, had they chosen to do so, its owners could have made considerable trouble for us, not so much by making us take the condemnation proceedings into court as by causing too many people to talk about what we were doing. When the initial overtures were made to them, I was most relieved to find that they were anxious to get rid of the school, for they had been experiencing great difficulty in obtaining suitable instructors since America had entered the war, and were very happy indeed to sell out to us and close down for the duration – and, as it turned[. . .
>
> (66)

Other scientists found New Mexico's landscape fascinating:

> "It was, Emilio Segrè writes, "beautiful and savage country": the dark Jemez Mountains to the west that formed the higher rim of the Jemez Caldera, the slumped cone of the old volcano of which Los Alamos was eroded tuffaceous spill; precipitously down from

the mesa eastward the valley of the Rio Grande, "hot and barren" except for the green meander of the river, writes Laura Fermi, with "sand, cacti, a few piñon trees hardly rising above the ground, and space, immense, transparent, with no fog or moisture"; farther east the wall of the Rocky Mountains as that range extends south into New Mexico to form the Sangre de Cristo, reversing hue from green to red progressively at sunset . . . "My two great loves are physics and desert country," Robert Oppenheimer had written a friend once; "it's a pity they can't be combined."

(Rhodes 451)

And Rhodes concludes: 'the Manhattan Project acquired its scenic laboratory site' (451). The 'beautiful and savage country' would host the most sophisticated technology of the era and produce the most savage weapon humankind had ever seen. The descriptions, the images and the framing context in Groves, Rhodes and other historians of the atomic bomb direct us to the atomic-industrial sublime well *before* the bomb itself. The aesthetically pleasing landscape would be transformed into a technological-industrial space. For the scientists themselves, as Joseph Masco has argued,

> the pleasures of nuclear production – of experimental success – have al- ways been mediated by the military context of nuclear explosions, requiring a complicated internal negotiation of the meaning of the technology . . . an experience of the nuclear sublime for weapons scientists, I would argue, is always an eminently political thing.
>
> (3)[4]

The atomic-industrial sublime employs a considerable quantitative rhetoric to communicate the scale of operations, most notably in terms of the consumption of resources and constructed structures such as pipelines, buildings and transportation. For instance, Rhodes writes:

> Wigner's team designed a 28- by 36-foot graphite cylinder lying on its side and penetrated through its entire length horizontally by more than a thousand aluminium tubes. Two hundred tons of uranium slugs the size of rolls of quarters would fill these tubes. Chain-reacting within 1,200 tons of graphite, the uranium would generate 250,000 kilowatts of heat; cooling water pumped through the aluminium tubes around the uranium slugs at the rate of 75,000 gallons per minute would dissipate that heat.
>
> (Rhodes 498)

Groves writes: 'an estimated 25,000 gallons per minute of water would be needed, assuming recirculation of cooling water' (71). Water restrained and channelized in such large quantities through pipes, tanks and holding areas, as Jim Moss would argue in his essay on the Three Gorges dam in China, is itself an instantiation of the sublime:

> Bound together by the alchemical wedding of its two crucially inter-dependent components, cement and fresh water, the resultant inert substance is the very antithesis of the fluidity inherent in the natural ebb and flow of age-old telluric currents . . .
>
> (252)

This merger of concrete, water and metal further underscores an aesthetic of the atomic-industrial sublime which anticipates the test and the bombings.

Other innovative modes of representing the atomic sublime occur in visual-verbal accounts such as Fetter-Vorn's *Trinity: A Graphic History of the First Atomic Bomb*. Here the emphasis is less on the industrial than on the science of the bomb.

We are given a brief history of the discovery of radioactivity, through the representation of Marie and Pierre Curie working in 'Paris, France, 1898' (Fetter-Vorm 4). The scene is a laboratory – the first of the man representations of laboratory space/life in *Trinity* – and the Curies are at a table carrying assorted bottles and devices, none of them distinguishable, being drawn as grey-black block objects. It is the Curies who set out to 'discover' the 'elements polonium and radium, which both emitted a mysterious energy'. Fetter-Vorm then *draws* the process of radioactive decay where U[ranium] becomes Th[orium], which becomes Ra[dium], which becomes Po[lonium] which becomes Pb (plumbum, lead). These elements are only represented by their periodic table symbols, each written in thick lettering with shadows:

U, Th, Po, Ra, Pb

(4)

We are told that 'radioactive decay . . . is happening all around us in nature'. The visual representation of the breakdown of Uranium into its end-element, lead, is *reduced* to the letters of the alphabet and the periodic table. The representation of what science believes takes place in nature are 'theories' of how the process occurs in nature (Rifkind 19). This decay, summarized as a visual equation, is what the scientists would seek to first replicate, then harness, in controlled conditions: conditions that would produce the atom bomb.

This representation of radioactive decay, incidentally, is not given to us in a panel, as is the case with many of the technical-theoretical aspects of the science of the atom bomb, almost as though the scientific process and its representation cannot be confined to the panel. On the facing page, for example, we are given, in sharp contrast to the lettered representation of radioactivity, the visual representation of the structure of the atom as discovered by Ernest Rutherford. The instantly recognizable planetary model of the atom depicts the nucleus with the electrons whizzing around it in fixed pathways, and the textbox adjacent to this image tells us: 'but it was the astoundingly powerful forces circulating through this nucleus that captivated Rutherford's attention' (5). Again, as in the case of radioactivity, this visual imaging of the atom is unbounded. Together, placed on the facing pages, the representation of radioactivity and the structure of the atom serve a key role in Fetter-Vorm's narrative: the energy of an atom and the atomic processes is literally and figuratively uncontainable. This interpretation of the image of the uncontainability and limitlessness of atomic energy as a potentially lethal force in Fetter-Vorm is invited by the stand-alone panel on page 5. Rutherford's face occupies the full panel with no vacant space (as opposed to the 'open plan' images of radioactivity and atomic structure). The textbox abutting into the Rutherford image says: 'to release those forces meant gaining access to a nearly limitless source of energy'. Rutherford has a speech bubble that says: 'some fool in a laboratory, if he finds the proper detonator, might just blow up the universe unawares' (Fetter-Vorm 5). Rutherford's words draw our attention away from the *previous* science drawing – of the process of radioactivity that 'is happening all around us in *nature*' – to the process that could be done in a *laboratory*. That if the previous image sought to inform us of a process that occurs in nature, Rutherford's quote directs us to the potentially dangerous harnessing, even if in the process of scientific inquiry, of these forces within a man-made, man-controlled environment: the laboratory.

The laboratory is a closed, even perhaps secret, space – secrecy, we know, was central to the Manhattan Project and to the transnational pursuit of nuclear energy itself – in which humans seek to unpack the processes of nature. Then, the laboratory is the space, as Svetlana Alpers argues, where 'the impact of the interference of the human observer in an account of natural phenomena' may be seen (Alpers 404). Rutherford is referring to the human observer, or catalyst, who, in the process of studying the phenomenon of atomic forces not in nature but in the *laboratory*, might 'just blow up the universe'. Taken together, the representation of the Curies' and Rutherford's *laboratories*, framed by and framing *natural* processes such as radioactive decay and atomic forces, respectively, on the facing pages (4 and 5), constitutes an interesting visual representation of the science of the atom bomb. Pages 8 and 9 have detailed visual images of the process of uranium fission. These are full-page images, essentially diagrammatic explanations of the process that is fission, except for five small panels. Within these panels, which serve as paratexts, or parergons to the fission process, Fetter-Vorm depicts: nuclear reactor chimneys, the radiation symbol, an X-rayed hand, a beaker with a liquid and a stirring rod. The juxtaposition gestures at the point Alpers makes: the natural decomposition or alteration of elements is engineered in laboratory/industrial conditions through human intervention. Science is the taking into humanized spaces a process that occurs out in nature, perhaps at an entirely different level or order. Thus, the inset panels and laboratories are not just the parergon, the outwork, to science: they *frame* the science as an interpretation of a process even as the natural process is instrumentalized *within* the laboratory. Later we are shown several sites of such an instrumentalization: the Universities of Chicago and California (Berkeley), Oak Ridge (Tennessee) and Hanford (Washington) where work on various aspects of the atom bomb is underway (20). There are other laboratory and industry images as well (36, 41–2, 46–7, 54–5, 57 and elsewhere). Similarly, there are many *unbounded*, non-panelled images of the fission reaction and other atomic processes (44–5, 48, 52–3, 58–9, 70). Most of these pages depict humans on the margins and sidelines, mostly set into panels, of the chemical process which itself, as noted before, is *not* empanelled.

The representation of the atomic processes and the humans who observe, explicate and eventually control these processes exhibited in Fetter-Vorm's visual rhetoric might be read in terms of what Peter Galison and Alexis Assmus have identified as the mimetic and the analytic experimentation modes (Galison and Assmus 1989). In the former, the experiment seeks to reproduce conditions and processes from nature in the laboratory; in the latter, the scientist seeks not to study the world reproduced but the *real* things that were made visible in the laboratory's conditions. I suggest that Fetter-Vorm's visual rhetoric depicts the interplay between these two forms of experimentation. The numerous references to the state of the atom are made to the state as it occurs in nature, even as these atoms are relocated into laboratory chambers and test tubes, indicating mimetic experimentation. However, the laboratory and the scientist are not interested in the decay into half-life of the elements, since they wish to see *now* the effect of the implosion of the atom: the energy released. This latter approximates to the analytic experimentation of Galison-Assmus because Oppenheimer and his cohort are not interested in the decay of the atom or its inherent instability, which is its natural state. The laboratory process mimes the decay of the atom in the natural world, but is interested in the *real* elements – polonium, thorium, isotopes of uranium – and energy, produced in the laboratory's miming of the process. The images and diagrams of the atomic processes – the science *of* the atom – framed by scientist-observers and laboratory settings – the science *around* the atom – oscillate between the mimetic and the analytic because they direct us simultaneously to a *fact* of natural decay, the *artifice* of reproducing it in the laboratory, and the explication of the real constituents – atomic energy – of the process. It is because of this key visual narrative strategy of locating, framing and embedding the atomic process within the human, laboratory, institutional settings that we recognize the mimetic and analytic dimensions of the 'science of the atom'.

Witnessing the Atomic Sublime

The distanced viewing of the atomic explosions, as recorded in the writings and images from the Trinity test by the scientists present there, and others, expound on the beauty and power of the explosions, on awe and wonder – and some horror – at what mankind has discovered and created. Many of the eyewitnesses remark on the sight in a rhetoric that merges the natural sublime with the technoscientific one.

William Laurence, the only journalist at the Trinity test, spoke in his *Dawn Over Zero* of the 'elemental creation of matter' revealing man's control over 'cosmic forces' (50–1), the 'bend[ing of] the forces of nature to his will' (51). He spoke of the event 'illuminating earth and sky' (22), the 'elemental flame' (22), a 'cosmic fire' (30), and a sound that seemed as though 'it came from some supramundane source as well as from the bowels of the earth' (24).

> It was like the grand finale of a mighty symphony of the elements, fascinating and terrifying, uplifting and crushing, ominous, devastating, full of great promise and great forebodings.
>
> (24)

He adds:

> It was a sunrise such as the world had never seen, a great green super-sun . . . as though the earth had opened and the skies had split.
>
> (31)

He compares it to the 'birth of the world' and the 'moment of creation' (32). Later, Laurence records the aerial sight of Nagasaki bombing:

> despite the fact that it was broad daylight in our cabin, all of us became aware of a giant flash that broke through the dark barrier of our ARC welder's lenses and flooded our cabin with an intense light . . .
>
> By the time our ship had made another turn in the direction of the atomic explosion the pillar of purple fire had reached the level of our altitude. Only about 45 seconds had passed. Awe-struck, we watched it shoot upward like a meteor coming from the earth instead of from outer space, becoming ever more alive as it climbed skyward through the white clouds. It was no longer smoke, or dust, or even a cloud of fire. It was a living thing, a new species of being, born right before our incredulous eyes.
>
> At one stage of its evolution, covering missions of years in terms of seconds, the entity assumed the form of a giant square totem pole, with its base about three miles long, tapering off to about a mile at the top. Its bottom was brown, its center was amber, its top white.

But it was a living totem pole, carved with many grotesque masks grimacing at the earth . . .

there came shooting out of the top a giant mushroom that increased the height of the pillar to a total of 45,000 feet. The mushroom top was even more alive than the pillar, seething and boiling in a white fury of creamy foam, sizzling upwards and then descending earthward, a thousand old faithful geysers rolled into one.

It kept struggling in an elemental fury, like a creature in the act of breaking the bonds that held it down. In a few seconds it had freed itself from its gigantic stem and floated upward with tremendous speed, its momentum carrying into the stratosphere to a height of about 60,000 feet.

(Laurence 21)

Laurence is awestruck by the power of the bomb, which he again (as in the case of Trinity) likens to the 'elemental' (which seems to be his favourite descriptor) power of nature, and does not speculate on the extent of destruction wrought by the bomb. The panoramic view of the explosion suffices as an instantiation of great power and beauty.

Other eyewitnesses too found the grandeur and magnitude of the phenomenon at Trinity breathtaking. Segrè said:

The most striking impression was that of an overwhelmingly bright light. . . . I was flabbergasted by the new spectacle. We saw the whole sky flash with unbelievable brightness in spite of the very dark glasses we wore. . . . I believe that for a moment I thought the explosion might set fire to the atmosphere and thus finish the earth, even though I knew that this was not possible.

(Rhodes 673)

That the sight and power defied any frame of comprehension the scientists and witnesses possessed is also well documented in accounts such as Rhodes':

"Most experiences in life can be comprehended by prior experiences," Norris Bradbury comments, "but the atom bomb did not fit into any preconceptions possessed by anybody."

(Rhodes 674)

Isidor Rabi said:

This power of nature which we had first understood it to be – well, there it was.

(Rhodes 675)

Mark Fiege has noted about these descriptions of 'nature's' power unleashed:

> Physical science opened infinite vistas on a range of phenomena, from the fantastically small to the incomprehensibly large; the scientists' realization of the vastness, unity, mystery, and sublimity of the cosmos evoked feelings of wonder that drove them onward . . . to the extent that the atomic scientists were able to describe and interpret their bizarre subject, they had to exercise a faculty more often associated with artists than with people such as themselves – the imagination'.
>
> (581–3)

Fiege also observes how the accounts emphasize the role of the settings of New Mexico – the desert, the mountains – were deemed as inspirational by the atomic scientists (see also Banco 2012).

In his analysis of the mushroom cloud images shot by the Americans over Hiroshima and Nagasaki, Peter Hales has argued that their sense of the sight was overawed by 'the absolute magnitude, the near-infinite power that the atom bomb represented, a phenomenon that set it apart from all that had preceded it' (8). Yet it was an 'abstract visuality' where 'none of these images showed the ground, the city, the target or the destruction' (9). The rhetoric, he notes, opted for analogies from nature and natural processes (as we have seen in Laurence's account):

> Most notable was the emphasis on natural imagery. By choosing such analogies, the writers did more than simply appropriate a language that could illuminate this new phenomenon. They bridged a previous gap between what was human and what was natural – the atom bomb became a man-made marvel of nature, and thereby the question of responsibility for the effects of the explosion remained slippery.
>
> (10)

Hiroshima in these pictures 'appeared not to be eradicated, but rather obscured by haze' (10). Citing Laurence's account Hales notes how he naturalizes a man-made event (11). Hales adds:

> Instead Laurence's sublime represented the furthest extreme of a twentieth century American version of the term, its translation from terror to tourism. Laurence's description introduced a new atomic aesthetic to Americans, one that converted holocaust to parlor show . . . and responsibility to mere response.
>
> (12)

Other than the images that Hales so perspicaciously examines, there are the ones that offer not the clouds but panoramic views of the cities.

Published by the Manhattan Engineer District under the title *Photographs of the Atomic Bombings of Hiroshima and Nagasaki* in three parts (1945), these panoramic shots – they are in fact titled 'panoramic views' – do not show even a haze. Almost indecipherable for any kind of detail, one sees only a vast expanse of grid-marked landscape. Some geological features such as rivers are discernible in the black-and-white images. Devoid of the emotional register of eyewitness accounts and placed, instead, within the register of geography and topography, these panoramas constitute an immersive experience which de-sublimates the sublime, converting the devastation into a mere geological/scientific map.[5] The entire view and emotion of the atomic sublime is, of course, made possible by the technoscientific achievement of flight: the humans view the cloud and the city from an aeroplane, distanced and safe from the destruction.

Now, historically the panorama, as Alison Griffiths informs us, 'laid claim to the historical and geographical real through an indexical bond, premised on their status as topographically correct and authentic reconstructions of battles, landscapes or ancient antiquities' (2). The moving canvas in the original nineteenth century panorama

> *reconstructed* a scene from history, a newspaper headline event or the natural world, it did not literally reenact this event for the spectator. Because the panorama was an image frozen in time, the scene was not literally reperformed for the spectator as in a film reenactment of a battle or an execution; there was no *action per se* in the painting.
>
> (3, emphasis in original)

The panorama was an 'artificial reality' that blurred the boundaries between the real and the synthetic, notes Griffiths (3–4). As an immersive experience, the panorama drew the spectator in through the reenactment of topographical detail but minus the actual events unfolding on the ground – this is precisely what the atomic sublime of the aerial panoramic views undertake.

This atomic sublime of a Hiroshima or Nagasaki, as constructed in the aerial panorama in the American photographs, relocates the spectator into the cities that are devastated but does not, in the photographs, show clear evidence of this devastation. Typical of the sublime's distanced view, the panoramic images are shot from elevated axial points that only serves to demonstrate the American power over the life of the city. As Paul Virilio argues, an aerial vision is inherently threatening for it hinges on an inhuman distance from those on the ground, especially those the vision and its accompaniments – gun fire, bombing – annihilates (Virilio 1984). In this case, the vision was located on either side of the destruction: the photographs of the cities are from 'before' and 'after' the bombing. It therefore gave the American viewer of these images of the cities, a near-totalizing perspective of distant (enemy) lands. The viewer experiences the cities virtually, like the Victorian spectator in the hot-air balloon (see Byerly 2008for a study).

An interesting variant of this form of the sublime where the expanse of destruction is widened may also be found in Hideo Furakawa's novel, *Horses, Horses, in the Innocence of Light*, that begins with the Fukushima disaster:

> Then, concentric circles. At first, an order for everyone within a three-kilometer radius of the Fukushima Daiichi Nuclear Power plant to evacuate; then, an order requiring everyone within a ten-kilometer radius to remain indoors. Before long the evacuation order was extended out to ten kilometers. An evacuation order was also mandated for everyone within a ten-kilometer radius of Fukushima Daini nuclear power plant; at the same time the evacuation zone was extended to twenty kilometers around Fukushima Daiichi. Two sets of concentric circles. In places they overlap. But, before long, a thirty-kilometer radius circle was added circling Fukushima Daiichi, inside which was required "internal refuge." This "big circle" looked like the corona around a sun. Around Daini was the "small circle." Subordinated to that "Big Circle" was a concentric core circle which made the Fukushima Daiichi Power Plant look like the sun. Land of the Sun. The new country of Japan.
>
> The concentric circles, all of them, lost their shapes, collapsed. Already on April 11 it was announced that the circles would be done away with "someday" . . .
>
> While these two circles, the big one and the small one, vie with each other, they are actually collapsed into one big circle, which results in the second "Land of the sun." The new Japan birthed from this is lumped together by name and geography with "Fukushima." The entire world associates it with this place. It became clear to me again. Fukushima Prefecture was being locked down; no, let's be precise: it was being blockaded.
>
> (24)

The circles are panopticon-like, although not perceived as such from the ground-level. But Furukama captures the sense of widening danger effectively through his verbal account. In *Chernobyl*, when Valery Legasov is told of the creation of a 30-kilometer 'exclusion zone', he is immediately furious. He asks:

> Someone decided that the evacuation zone should be 30 kms when we know cesium 137 . . . is found 200 kms away – an uninformed, arbitrary decision that would cost . . . how many lives.
>
> (episode 3)

The panoramic image retains a certain cartographic appeal reminiscent of the traditional panorama but also suggests, when we look down at the photograph in print form, a vertiginous dizziness of viewing and interpretation. Any humans on the ground in the cities have been miniaturized to the point of being invisible. This, then, is a *sanitized* atomic sublime of destruction, devoid of the presence of loss and death.

The Catastrophic Sublime

Nuclear catastrophe, Jean-Luc Nancy points out, is incalculable in its effects. Writing in the immediate aftermath of Fukushima, Nancy says:

> No one can truly calculate the consequences of Fukushima, for humans, for the region, the earth, the streams, and the sea, for the energy economy of Japan, for calling into question, abandoning, or increasing control of nuclear reactors all over the world, and thus for the energy economy worldwide. But all this is incalculable because it challenges the capacities of calculation whereas, at the same time, what we plan or project remains within the order of calculation, even if it is out of our reach.
>
> (27)

Nancy is signalling the sublime of incalculability.

If the aesthetic of the atomic sublime in American accounts and images of the tests and the bombed cities attempted to focus on the power of nuclear energy, a different order of the sublime emerges in images that capture the devastation at close(r) quarters. In texts such as *Barefoot Gen*, gruesome scenes of injured bodies, destroyed houses and close-ups of the injuries themselves suggest an aesthetic perspective – Hillary Chute speaks of the 'grotesque clarity and directness' of *Barefoot Gen* (126) – that seeks to make visible the trauma of war in unprecedented ways. The catastrophic sublime is a sublimity of collapse, of cities razed to the ground and evoking a sense of being distressed rather than uplifted (the latter being a feature of the traditional sublime). It is an aesthetic that tries to capture the incalculable.

The Decadent Sublime

After a series of aerial photographs of the Nagasaki area, Donald Goldstein et al. in *Rain of Ruin: A Photographic History of Hiroshima and Nagasaki* (1999) offer what they call a set of two panoramas: 'a panorama of four photos taken from ground zero, west at center' and 'a panorama of four photos taken from ground zero, northeast at center' (92–3). This is a panorama that enables us the spectator/reader to perceive the destruction at closer range. As befits the term and the original, nineteenth-century technique of sprawling canvases, Goldstein et al.

arrange these photographs across a two-page spread. When we view these pages, the images surround us from all sides and multiple angles, generating an immersive experience of the catastrophic sublime as the previous aerial shots did not.

The arrangement of photographs in these volumes is also interesting as they attempt to provide a centrifugal sense of the catastrophe. The images are arranged in terms of the scenes moving outward from ground zero. Thus, we have images from '7500 feet from the hypocenter' (Goldstein et al. 89), 'from near ground zero' (Goldstein et al. 94), '1000 feet out' (Goldstein et al. 94), '1/4 mile from the center of the blast' (Goldstein et al. 95), '14, 125 feet southwest of ground zero' (107) and so on. Forcing us through these numerically organized and spatially calibrated photographic textual framing, the collections of photographs invoke what Immanuel Kant identified as a mathematical sublime which 'which begins with the inability to comprehend the scale and vastness of a mountain or a river' (Masco 3) and where magnitude is *attempted* to be apprehended (as opposed to the aesthetic mode of comprehension) by means of numerical concepts. Apprehension here is the tension between the imagination of many individual parts and the sense of the whole where these parts are unified.

What the photographs of such collections achieve via this mathematical sublime, a subset of the catastrophic, is an attempt to communicate the vastness of disaster in terms of a rationalizing process which, ironically, only points to the immeasurable magnitude of devastation. That is, if the mathematical sublime 'also arises from revitalization by our reason, since the infinite or indefinitely plural "can nevertheless be *thought*" as a unified whole' (Hart 842, emphasis in original). The revitalization does *not* provide succour: on the contrary, and in line with Kant's theorization, it is grounded in despair, literally, even as we seek a conceptual unity of the multiple zones, scenes and demographics of devastation. Emphatically, this is *not* a safely distanced panoramic view: it is a despairing sublime that emerges from the images where the highlighting of distances from ground zero only spreads the devastation, literally, across landscape, vision and our apprehension.

Other Japanese photographs from ground zero outwards capture the devastation in terms of human losses amidst the ruins. Images of destruction include the ruins of houses, buildings, urban infrastructure and the landscape. Yamahata captured a ruined wagon in Nagasaki and Satsuo Nakata, the remnants of a tram car in Hiroshima (*Flash of Light, Wall of Fire*, 36–7). Masami Oki shot a photograph 'Remains of a large house', 20 August 1945, in Hiroshima (*Flash of Light, Wall of Fire*, 26–7).

Vast landscapes of destruction with the detritus of some building are instances of a 'decadent sublime', captured through the gritty, amateurish and often flawed photographs and narratives in the immediate days and months after the bombings. Jennifer Presto, theorizing this version of the sublime in images/representations of massive destruction, such as earthquakes, argues that the decadent sublime 'reverses the upward vector of the individual's cognitive response that is central to Kant's dynamic sublime and, for that matter, the very definition of the sublime' (577). Instead of rising, the human falls 'back to earth in a downward vector that is decadent in the primary sense of the word, which is derived from the Latin *de* and *cadere* meaning "to fall down" ' (578). The falling-down is about the ruins of structure and the emptying of space.

This sense of being overwhelmed by the sight of mass suffering and destruction is captured in Yōsuke Yamahata's comments about how he took those photographs of Nagasaki. Yamahata said:

> Perhaps it's unforgivable, but in fact at the time, I was completely calm and composed. In other words, perhaps it was just too much, too enormous to absorb.
>
> (Cited in Marcoń 794)

But absorb he did, and aestheticized the suffering too, as Marcoń observes. Yamahata's comment about the sights being 'too enormous to absorb' and his sense of emotional detachment ('calm and composed') captures the (sublime's) sense of demolished, or at the very least diminished, self in the face of massive sights of suffering and threat. As Tom Cochrane puts it:

> the sense of self-negation is a sense how physically insignificant, or utterly contingent we are in comparison to the object. And this, I claim, is a necessary component of the sublime experience. Although the experience is focused primarily on the object, to see something as big or powerful is at the same moment to feel small and vulnerable. Even when looking at the landscape from the top of the mountain, one may feel reduced by the magnitude of the earth. It is the feeling that comes from confronting something inhuman, uncompromising, hostile or just profoundly indifferent. And this can be grasped in a single perceptual experience that startles or overwhelms the spectator, or it can emerge more slowly in contemplation.
>
> (130)

The photographer on the scene was caught between wanting to capture the devastation and a sense of self-preservation: 'Yamahata wrote that he thought only about himself, how to take pictures and to avoid death if another attack occurred' (Marcoń 794). Here, the catastrophic sublime is manifest in the implicit sense of the loss of the self in the face of the massive disaster.

Tetsuo Miyata captures the seeming endlessness of the devastation when he recalls how, as a child on the day of the bombing, he walked through Hiroshima:

Lines of wounded stretched unbroken on both sides of the road . . .
Over the endless expanse of rubble rise the coal-black remnants of
trees.

(Osada, unpaginated)

This loss of self and the disorientation of viewing vast acres of land bereft
of people – the photographs of devastated landscapes in collections such
as *Rain of Ruin* are often prefaced by photographs of the cities *before*
the bombing – is also signalled, and generated, through the depopulation
narrative that the images set up. From the memories (for, say, Yamahata)
or the views (for us readers) of the once-populous city, the post-bombing
photographs invoke a decadent sublime through the depopulation
narrative encapsulated in the images and descriptions.

One survivor reports:

Hiroshima had disappeared . . . that experience looking down and
finding nothing left of Hiroshima-was so shocking that I simply can't
express what I felt. . . . Hiroshima didn't exist-that was mainly what
I saw-Hiroshima just didn't exist.

(Kyoko and Selden xx)

In *Barefoot Gen*, Nakoake's response to the sight of Hiroshima after the
bombing is 'everything is gone, everything has disappeared' (II: 7) and
full-page spreads show the extent of destruction, with buildings labelled
for us to know what exactly has been lost (II: 8–9). Panels zero in on the
collapsed houses, with such panels repeating across the volumes as Gen
recounts not only the destruction of his home but that of others as well,
once again underscoring the *scale* of destruction and loss (I: 254–5 and
elsewhere). Most shots in *Chernobyl* are of empty houses, streets and
the Exclusion Zone itself. There are also shots of houses abandoned in a
hurry, lending an eerie effect (episode 2).

Wilfred Burchett speaks of how he

picked [his] way to a shack used as a temporary police headquarters
in the middle of the vanished city.

(2)

Then there are the images Yamahata, Onuka and others captured of the
dead and injured humans, constituting a horrific archive in a different
register. Masama Onuka, Yōsuke Yamahata and Eiichi Matsumoto shot
close-up pictures of burnt bodies. Where Japanese army personnel are
photographed, they are less heroic than figures of pathos (the photographs
by Shunkichi Kikuchi, *Flash of Light*, 82–5).[6]

The decadent and catastrophic sublime is an aesthetic of massive
suffering seen up close, and therefore uncomfortably proximate. In the

catastrophic sublime, the photographs demonstrate and capture the micro-manifestation of a terribly sublime power: the sublime impinges and impacts the bodies of the people. The burns, injuries and scar tissue on numerous bodies and, in later accounts of radiation sickness, is toxicity writ large.[7] They are material witnesses to the violence of the war and the bombing. (During the Occupation, the Japanese accounts and photographs were censored and banned – so many of these were hidden away, and saw light of day years later. The first colour footage of the destruction, taken by Herbert Sussan, was classified as top secret by the Pentagon, who, 'did not want Americans to see graphic evidence of the bomb's effects on human beings'. The footage finally started appearing in American films such as *The Day After* and *Dark Circle*, in the 1980s (Lifton and Mitchell 259–61).

There is another angle to the photographic representation of the ruined landscapes of Hiroshima and Nagasaki, to which I now turn. In the Cold War period, these images were circulated by the American government to indicate the extent of damage that may result from American cities being attacked. These images, then, fuelled the call for civil defense and campaigns for nuclearization. Joseph Masco writes:

> The immediate project of the nuclear state was thus to calibrate the image of atomic warfare for the Ameri- can public through the mass circulation of certain images of the bomb and the censorship of all others. In this way, officials sought to mobilize the power of mass media to transform nuclear attack from an unthinkable apocalypse into an opportunity for psychological self-management, civic responsibility, and, ultimately, governance.
>
> (257)[8]

The Japanese cities' ruins, then, are reappropriated for an entirely different purpose in the present: by the Americans.

The injured, burnt and dead bodies enact in the flesh, the catastrophic sublime. That is, the sublime is materialized in the form of the broken, burnt and damaged bodies as captured in the photographs and, as we shall see, the paintings. Such a sublime is not about the safety of distance. Rather, it is the sublime of horrific proximity, generated through the sights of numerous injured bodies, when we stare at series after series of them. While such a sublime is undeniably linked to trauma (in terms of the sensory assault on the viewer), it serves as a rehumanizing response to the distanced sublime of aerial surveys. This is not a step-back from the abyss or terror: it is sublime by virtue of the photographer, artist and viewer being *immersed* in it and, as a result, evoking/experiencing strong emotions.

In artists seeking to render the excessive suffering from the bombs, writes Hillary Chute about Nakazawa and *The Hiroshima Panels*, among other texts:

The disjuncture, or lack of disjuncture, between the 'exaggerated' rendering in the story – much of which is conventional to manga – and the real, decimating violence of the bomb throws into even greater proportion the catastrophe of 'the real' in this narrative.

(126)

The construction of hell in the Marukis' *Hiroshima Murals* (15 paintings, first exhibited in 1950), as Charlotte Eubanks (2009) and John Dower (1985) have argued, recalls the events through multiple angles:

The mostly monochrome ink washes show tangled masses of naked human bodies punctuated with the occasional red of blood or flame. Each mural explores the aftermath from a different iconic angle: the procession of victims away from the hypocenter, people huddled in bamboo thickets for shelter, those who drank the poisoned water and died, those burned in the secondary fires that scorched the city.

(Eubanks 1614)[9]

Eubanks notes how 'the human body in pain [is] a central motif' in the murals (1618). Echoing Eubanks, Richard Minear, in his review of the collection, *The Hiroshima Panels* (1985), declares: 'the primary testimony of the Marukis is to the human impact of the atomic bomb' (59), and speaks of the experience of viewing the full-scale murals in the gallery, which 'overwhelm the viewer' (60). Minear also notes that 'the paintings focus so completely on people that Hiroshima is not recognizable in the "atomic-bomb paintings"' (63). The cities after the bombings *was* hell, as recorded by numerous hibakushas, including those who were children at the time (see *Children of the A-Bomb: Testament of the Boys and Girls of Hiroshima* for examples of such descriptions).[10]

This shift from Hiroshima as the locus and focus of the work to the human as the 'site' of war's horror instantiates a different order of the sublime.

The Maruki murals are, as commentators note, centered on bodies of suffering, never allowing us the luxury of moving away from the corporeal sites of war. The realism of the paintings, especially of the 'humans in a group' embodied a collective memory, argues Yukinori Okamura (522).

Unlike the photographs of devastated landscapes of identifiable/ identified cities (Hiroshima, Nagasaki) that are more or less emptied of people, and which invoke their own version of the sublime, the Marukis by erasing all contextual material in the diegetic space of the mural (although the textual commentaries mention the cities), universalize the irradiated human form. The 'atomic desert', then, is any and every place that has irradiated humans, dead or dying. (When in later murals they include Auschwitz, the Minamata disaster, Bikini Atoll and American PoWs in Japan, they of course expand the frames of reference beyond Japanese cities.) The violence appears in murals like 'Fire' in the form of the red blaze that envelopes the human form, invoking images of hell, as Eubanks notes. The mass of skeletons and bodies, both revolt and fascinate. Expressions of grief and suffering in murals like 'Water' evoke *universalized* states of affect in situations of extreme suffering.[11] In the midst of mass dehumanization wherein entire populations were wiped out in the first moments of the blast, the Marukis call attention to the individual bodies – even when massed – that make up these eradicated populations. They refer to the 'smell of corpses [that] hung in the wind' (51). In cases such as the inversion of the Madonna-and-child symbol, wherein the mother feeding her child discovers the child dead at her breast, as the text-commentary tells us, the Marukis write:

> An injured mother cradling her dead infant. Is this not an image of despair? Mother and child should be a symbol of hope.
>
> (39)

In the place of the respected and respectful interring of the bodies, the separation of the bodies of the living and the dead, the Marukis offer us masses. In their text to 'Bamboo Grove', which depicts the homeless seeking refuge in the grove, and dying there, they write:

> No one disposed off these corpses, and they were not moved until a typhoon in September washed them out to sea.
>
> (Dower and Junkerman 55)

The dead and the dying fuse in the large masses that they drew. The accompanying text often underscores the indistinguishability of the dead and the dying, for example, in 'Water':

> There were mountains of corpses, piled with heads at the center of the mound. They were stacked so their eyes, mouths, and noses could be seen as little as possible.
>   In one mound a man's eyeball moved and stared. Was he still alive? Or had a maggot moved his dead eye?
>
> (Dower and Junkerman 39)

The text for 'Atomic Desert' reads:

> Even today, human bones are sometimes unearthed in Hiroshima.
>
> (51)

Commentators have detected a certain distancing in the Marukis' work, but it is not always possible that we can discern this distancing, given the intensity with which suffering has been portrayed, even if these images are later recreations and *not* witnessed scenes. The bones in the Marukis' work closely resemble the heaps of skeletons and skulls that abound in

Barefoot Gen. We can, then, think in terms of what Hillary Chute has called 'a collective idiom of witness', which generates an aesthetic of the catastrophe (Chute 130).[12]

While practices for the recently dead and commemorative ceremonies are culture-specific, the emphasis on the unidentified masses of the dead in the Marukis, suggests that 'the unnamed body, the universal unnamed body, representing all bodies that had lost their names, struck powerful chords of sentiment', as Thomas Laqueur puts it in his cultural history of human remains (though he is speaking of the war-dead and the efforts to create memorials, and memories, for the 'unknown' soldier, 481).[13]

But this is not all. The Marukis speak of discrimination even with the dead. In the mural 'Crows', they draw crows swooping down on and consuming corpses. They cite Michiko Ishimure who documented how crows descended on the unburied Korean corpses and ate their eyeballs. The accompanying text says:

> The Koreans were discriminated against, even in death. The Japanese discriminated, even against corpses. Both were Asian victims of the bomb.
>
> (83)

In *Barefoot Gen*, Mr Pak, the Korean neighbour of the Nakaokas who had helped them through the war, is transformed into an angry man when Gen encounters him days after the bombing. When Gen queries him, Pak tells him that his old father survived the bombing. But when he was taken to the hospital for burns, the Japanese doctors refused to treat him because he was Korean. His father eventually dies because he does not receive treatment (II: 167–8). Exposing the hypocrisy and internal racisms of the Asians, the Marukis universalize the dead in their catastrophic sublime that centres around the universalizable and universal human body.

Suzannah Blernoff argues about images of bodily pain and suffering that these serve as 'points of affective intensity' (65) evoking 'disgust and fascination, fear and arousal' – an ambivalence characteristic of the sublime (65). For Blernoff, following the work of Paul Crowther, the emphasis on corporeal suffering in images of war and destruction, has a redemptive element: a dehumanization in the face of the constant threat of the loss of the self (68). In contrast to the 'uplifting' sense in the face of the traditional/transcendental sublime and more aligned with the decadent sublime, Blernoff argues for a sublime that is unrelentingly corporeal, where grief and other strong emotions are represented (in the images) *in* the corporeal (70–1). Blernoff elaborates on this 'corporeal sublime':

> The feeling of the sublime occurs at precisely these thresholds-between self and other, subject and object, human and non-human. The sublime is fundamentally about frontiers. Where do we, individually and

socially, locate our limits: the limits of embodiment, self, civilization, humanity, meaning?

(71–2)

This sublimity of limits is precisely what the Marukis achieve in their merging of the human with the non-human, subject and object, living and non-living.[14]

The massing of the dead was also the result of careless and wilfully indifferent work by the American Occupation forces in Nagasaki:

> When American troops built an airstrip in the northwest corner of the valley – nicknamed Atomic Field – they used bulldozers to clear the ruins, crushing human bones scattered in the debris. "There were still many dead under the rubbish," fifteen-year-old Uchida Tsukasa remembered. "Despite that, the Americans drove their bulldozers very fast, treating the bones of the dead just the same as sand or soil. They carried the soil to lower places and used it to broaden roads there." . . . people who lived in the area, and those whose family members' bones were buried in the debris, could only stand by, outraged and helpless.
>
> (Southard, unpaginated)

Other instantiations of this catastrophic sublime in the form of non-living and non-human material witnesses also figure in the narratives around Hiroshima, Nagasaki and Chernobyl. The now-famous image of the shadowy outlines of a man on the stone step in *Flash of Light* (125) and the opening photograph of the volume – a clock stopped at 8.10, 'at the time of the bomb blast' (16–17) – are examples of material witnesses.

> *Material witnesses* are nonhuman entities and machinic ecologies that archive their complex interactions with the world, producing ontological transformations and informatic dispositions that can be forensically decoded and reassembled back into a history. *Material witnesses* operate as double agents: harboring direct evidence of events as well as providing circumstantial evidence of the interlocutory methods and epistemic frameworks whereby such matter comes to be consequential. *Material witness* is, in effect, a Möbius-like concept that continually twists between divulging "evidence of the event" and exposing the "event of evidence."
>
> (Schuppli 3, emphasis in original)

Technical objects, writes Schuppli, 'can account for and express their historical conditions; that artifacts can induce the affective register of testimony; and that materials can, in short, bear witness' (14). In many photographs from Hiroshima and Nagasaki, we come across burnt bridges, shattered buildings, human remains, abandoned submarines and biologists collecting earthworms at ground zero. When these objects are placed within the archives or volumes, they are made to tell the story of the bombings.

> Matter becomes a *material witness* only when the complex histories entangled within objects are unfolded, transformed into legible formats, and offered up for public consideration and debate.
>
> (Schuppli 18, emphasis in original)

In the catastrophic sublime of nuclear destruction, where the human and the non-human, the living and the non-living have fused, as we have noted in the Murakis and the photographs of entangled 'things', the material witness occupies the *same* affectively invested space as the oral narratives and written accounts of the *hibakushas*.

Or, the irradiated materials, including bodies both dead and injured, spread far and wide in terms of how they impinge on the consciousness of the survivors. Writes Hiroko Takanashi in 'Red Parasol', her account of the bombing:

> I still remember the smell of the atomic bomb, or perhaps it is more accurate to call it the smell of Hiroshima . . . the dead bodies quickly rotted and were infested by maggots . . . the smell of the discharge from those wounds, the smell of burning bodies and the smell of the burning city all mixed together and clung to our clothes.
>
> (Mizue 20)

Other examples of the intolerable smell that hung in the air occur in Kyoko and Selden (ix, 35). This is the materiality of many bodies that affects the senses of the survivors.

The bottled specimens in the ABCC (*Barefoot Gen*, V: 206) and the racialized politics of scientific research (already signalled in the chapter on 'nuclear subjects') serve to highlight the catastrophic sublime's gothic horror and its political sources. The specimens, test-cases and dead bodies – all Japanese – gathered by the ABCC are akin to the 'freaky' bodies that Rosemary Garland Thomson writes of:

> The physically disabled body becomes a repository for social anxieties about such troubling concerns as vulnerability, control, and identity. In other words, I want to move disability from the realm of medicine into that of political minorities . . .
>
> (6)

To return to an argument I made before in connection with the Bhopal disaster victims, the specimens in ABCC as depicted in works such as *Barefoot Gen* are 'a corporeal instantiation of a series of processes: the poisoning of bodies by toxins from M[ethyl] I[so] C[yanate], the absence of accurate medical knowledge, the lack of support structure and finally, poverty' (Nayar 101). The horror is literally distilled into bottles, and the bottles seek to *contain* the sublimity of destruction, a destruction that is unparalleled in scope and intensity.

In the case of Chernobyl, however, the carefully arranged set pieces of the Prypjat town that we encounter in the photographs of, say, *Chernobyl: A Stalker's Guide* offers a cinematic transformation of the material witness, a kind of rematerialization and resymbolization of the events through the choreography of objects. Darmon Richter writes:

> Gas masks dangled from light fittings. Dolls and other toys sat propped up in creepy poses on wireframe beds, like B-rated horror movie props.
>
> (54)

The whole effect, he says, was 'inauthentic' (54).

The post-disaster sublime, however, is a catastrophic sublime of a different order of material witnessing. Long shots of the abandoned town with the dome of the power plant's new arch in the distance juxtaposed with the Reactor Block 4 with its new protective shelters (104–5), the control room with the original switches and keys – 'now stripped of many of its fittings and cleaned of dust . . . declared safe for visitors' (204) – and the long pipelines around the complex (199) reinscribe the disaster site within a new economy of not only tourism but also continuing investment in nuclear power. Richter records that 'cracks in the shelter structure . . . releasing radiation from within' have been observed (212), indicating that the place remained dangerous. The icons – empty buildings, new buildings, material objects such as control systems and toys – are material witnesses whose function, whether in Hiroshima or Chernobyl, appears to be one of forging continuities.

In places such as Chernobyl, one encounters landscapes that are irreducibly toxic, just as Nevada, the Bikini Atoll, the Australian desert are toxic landscapes. Speaking about Chernobyl, Richter notes how, under the New Arch, lie materials that have half-lives of 246,000 years, 700 million years and 4.5 billion years (216), so that questions of 'containment' are only partially answerable when we think of the quantum of time involved in the full decay of radioactive materials. Tatsuta stares out at the ruined city of Ishinomaki and admits 'the scale [of destruction] here is stunning' (236). A 'mountain of rubble' is what he sees (237), and the images he draws (51–2, 96, and elsewhere) reminds one of the Edward Burtynsky photographs of landscapes of waste.

The landscapes that Peter Goin captures are, likewise, of massive acres of contaminated land, storage buildings and containers for radioactive wastes and abandoned nuclear plants are instantiations of a 'toxic sublime' (Peeples). Accompanying these photographs, drawings and paintings of toxic sublime landscapes are assessments of continuing radiation levels.

The ordering and transformation of the events of 1945 or 1986 occurs through the indexing of radiation-measurement and the radiation effects in material (organic) bodies that the narratives round nuclear disaster encode. As late as the 1980s texts on the nuclear legacy point to indexes such as radiation dosage absorbed by down winders in Nevada and other places, measuring cause and effect. Thus, Kirk and Purcell in *Doom Towns* note the inquiries into the tests and the quantum of radiation likely absorbed by the residents (124–5). Thus, the Marukis bring the narrative all the way to the Minamata disaster (mercury poisoning from the Sanrizuka plant in the mid-1950s) showing the ataxia in the victims and massed bodies (102–3). Richter notes that in Chernobyl's 'Exclusion Zone', nearly 30 years after the accident, the radiation level is 600 uSv/hr (microSievert/hour), a 'potentially dangerous dosimeter reading' (238). Goin notes in the text accompanying his photograph 'Ground Zero and Tower', that 'the land in this area is still contaminated with alpha particles' (36). In *ICHI-F*, Tatsuta describes how every day, as a clean-up worker in the Fukushima reactor premises, they had to adhere strictly to radiation check norms, every single day, on entering and exiting the plant (245–6, 261–2). He notes:

> Just 30 minutes of work gave me a dosage a full digit higher than a day at the shelter.
>
> (262)

The indexing of danger through, say, radioactivity measured by dosimeters carried by tourists into Chernobyl's Exclusion Zone, implies this continuity from 1986 to the present, just as the material bodies of *hibakushas* in their narratives mark a continuity: and hence the insistence I make on the transhistorical nature of the catastrophic sublime in the nuclear texts of the twentieth century. Such lists, rankings and ratings (of radiation and therefore of danger) connect New Mexico with Nevada, Hiroshima, Chernobyl and Fukushima.[15] In the process, these modes of indexing also point to the nuclear sublime as a state of permanent crisis. It is not bound in space (Nevada or Ukraine or Russia or Australia or Japan) or time (1945, 1986, 2011) but stretches across time frames and spaces when the gases, toxins, strontium clouds and radiation move through generations (in terms of effects passed on from one generation to the next, as seen in the case of the Japanese victims and down winders) as well as across geographical regions (the gases and toxins poisoning oceans and air across Europe, all the seas beyond Marshall Islands, etc.).[16]

There is one more aspect of the catastrophic sublime that merits attention: this is the transformation of the site of the bombings into a sort of research laboratory by the American scientists and doctors, who collected samples and materials from the *hibakusha* but refused medical treatment.[17] The Atomic Bomb Casualty Commission (ABCC) which led the research initiative followed a strict 'no-treatment policy' (Lindee 1994 Chapter 7). Susan Southard writes:

> the ABCC conducted medical examinations without also offering medical care. What Do-oh and other hibakusha didn't know was that the ABCC's mission to conduct detailed studies of survivors' radiation-related illnesses included a strict mandate to provide them no medical treatment.
>
> (unpaginated)

Norman Cousins, the journalist who would later initiate the 'moral adoption' campaign for the *hibakushas*, wrote scathingly of the

> strange spectacle of a man suffering from [radiation] sickness getting thousands of dollars' worth of analysis but not one cent of treatment from the Commission.
>
> (cited in Southard, unpaginated)[18]

Japanese doctors like Nishimori Issei expressed their concern about

> the ABCC's way of doing research seemed to us full of secrets. We Japanese doctors thought it went against common sense. A doctor who finds something new while conducting research is obligated to make it public for the benefit of all human beings.
>
> (cited in Southard, unpaginated)

Southard records:

> Many survivors hated being studied by doctors from the country that had irradiated them. The ABCC also transgressed cultural boundaries with invasive and intimidating procedures, by examining young people like Do-oh in the nude, collecting blood and semen samples, and taking photographs of survivors' atomic bomb injuries . . . Even the word "examination" seemed objectifying to many.
>
> (unpaginated)

In *Barefoot Gen*, the Nakaoke boys take their ailing mother to the ABCC. There she is stripped and 'checked every corner of her body. That was it' (V: 182). The boys were given a receipt titled 'Specimen Collection Date'. Gen's brother, Koji, is horrified at the receipt. The panel shows Koji glaring at the receipt and the speech box says:

> Specimen? They intend to use her as a specimen in some experiment, like an insect!
>
> (183)

After returning home, Koji tells his brothers:

> The ABCC sees the bomb survivors as nothing more than bugs under
> a microscope. I bet they're taking students from your school to serve
> as guinea pigs too.
>
> (183)

Gen and Koji realize that the horror of the post-bombing years is about
the absolute indifference of the ABCC and the Americans to Japanese
suffering:

> First they drop the bomb on us and kill our father
>   And then we let them use our mother as a guinea pig for their
> experiments.
>
> (184)

The catastrophic sublime and its horrific effects – including the emotional
trauma of the *hibakushas* – are compounded by a transformation effected
by the ABCC's operations, as Nakazawa captures in his account.[19] Beyond
the secrecy and treatment-denial was the transformation of the victims
of the bombings into medical source-material. Sample collections and
'examinations' as the order of the day effectively relegated the Japanese
victims into raw material for their (Americans') scientific studies and reportage. That is, Hiroshima and Nagasaki served as large-scale laboratories where specimen and spectacle came together in the massive number
of injured and radiation-affected bodies. Just as

> the battlefield over the last century became a crucial field laboratory
> [and] the United States did not bomb Hiroshima and Nagasaki, for
> example, as a scientific test. But the two cities became a scientific
> resource after the fact, for studies of both the biological effects of
> radiation, and of the physical damage produced by nuclear weapons.
>
> (Lindee 9)

In other words, the bombing was instrumental in transforming the cities
into not only a site of devastation and injured bodies but also, perversely,
these bodies into raw materials, further dehumanizing them and making
a spectacle and specimen of their wounds and ailments.

Temporality and the Catastrophic Sublime

Adam Higginbotham writes in *Midnight in Chernobyl*:

> In Kiev, even two years after the accident, young couples were afraid
> to have children, and people ascribed every kind of minor illness to
> the effects of radiation.
>
> (326)

And

> In February 1989, almost three years after the accident, a prime-time report on Vremya revealed to the Soviet people that the true extent of radioactive contamination beyond the perimeter of the thirty-kilometer Exclusion Zone had been covered up – and that the total area of contamination outside the zone was, in fact, even larger than that within it.
>
> (328–9)

Keiji Nakazawa's Gen asks whether the Japanese are condemned to remain *forever* as radiation victims and suffer endlessly. Kate Brown writes about Chernobyl:

> Chernobyl was not a single event but was instead a point on a continuum; the radioactive contamination of Polesia lasted more than three decades. Chernobyl territory was already saturated with radioactive isotopes from atomic bomb tests before architects drew up plans for the nuclear power plant. And, after Chernobyl as before Chernobyl, the drumbeat of nuclear accidents continued at two dozen other Ukrainian nuclear power installations and missile sites. Sixty-six nuclear accidents occurred in Ukraine alone in the year after Chernobyl blew. More nuclear mishaps transpired after the Soviet Union collapsed, including the fires in the Red Forest in 2017.
>
> (unpaginated)

In these accounts, the disaster that is the atomic bomb or the nuclear explosion is not just an event inscribed in time and space: it sediments in the soil, water, plant life, the DNA of the living and memories. Nakazawa's Gen, after the American occupying forces have torn down their house, stabs himself with a nail. To his horrified friends Gen says holding up his bleeding hand:

> I'm never gonna forget this pain . . . I'm never gonna forget this anger . . . If I start to forget what happened today, I'll look at this scar and remember.
>
> (IX: 30–1)

Throughout *Barefoot Gen*, panels often work as montages, 'proliferating temporalities . . . holding time in suspension' (Chute 73). Hillary Chute also points to the 'temporal languidity' of Nakazawa's art where he draws 'slow hordes of burned people who move dazedly . . . marching onward slowly and automatically' (124). What is significant is that this memory and image repeats throughout the volumes, thereby suggesting a sedimentation of the memories of disaster, their repetition and immanence, constituting the dwelling in the end time.

In other words, the reiteration of the injuries in memory *and* in representations, such as the ones Chute identifies in Nakazawa's 'temporal languidity', directs our attention to the timeless in the nuclear sublime, a permanent abject state of being in the end period. The nuclear sublime, as Nakazawa depicts it and as Higginbotham records of the Chernobyl survivors, is the unconscious itself being in a state of immanent apocalypse. As Patrick Wright argues about the endlessness of the sublime:

> Despite a diversity of affective responses, though, the tumultuous sensation of sublimity is accompanied by allusions to the "eternal," "timeless" and "infinite" . . . which are not simply terms in a mystical lexicon but also attest to the atemporality that is a mark of the unconscious drives and the abject.
>
> (Wright 91)

The abject is the very condition of repetition, in memories of those like Tetsuo Miyata:

> They had the air both of people who were asking for nothing, and again of those who were brimming over with an infinite petition.
>
> (Osada, unpaginated)

Miyata (and numerous others in Osada's collection of children's memories of the atomic bombing), Nakazawa's Gen exemplify one specific quality of the sublime: the role of the imagination in reproducing memories. Rudolf Makkreel in his comments on Kant's sublime writes:

> All representations are given successively in accordance with time (the form of inner sense), and their association in a series could occur without the power of the imagination to reproduce past representations in the present . . . The imagination plays a more important role in producing a complete reproduction.
>
> (305)

Makkreel adds:

> All imaginative reproduction would be in vain if we did not consciously recognize that what is reproduced in the present is identical with what was apprehended in the past.
>
> (306)

When Nakazawa repeats the drawings – they are exactly reproduced from volume I – of the burnt bodies of Hiroshima's residents, and the dying bodies of his parents, he is imaginatively reproducing in his present a continuing trauma. In the process, he also apprehends the past as a determining feature of his – and others' – present. The vastness of the devastation he witnessed can only be visualized as an iterative sublime.

There is a second sense in which the catastrophic sublime invokes scrambled and iterative temporality: a lifetime of radiation sickness, and the intergenerational nature of radiation poisoning. The intergenerational nature of radiation's effects resonates with the apocalyptic sublime's key feature: repetition and endlessness:

> The apocalyptic sublime erases teleology and substitutes traditional "ends" with the sublime experience; this experience attempts to post-pone resolution and even perpetuates the disorientation and anomie experienced by audiences, ultimately in order to destabilize the subject.
>
> (Gunn and Beard 284)

In Chapter 2, I had already cited accounts from the Utah 'Down Winders' where people like Victoria Burgess record:

> By our fortieth class reunion in 2003, one third of our classmates had died from radiation poisoning and various related diseases as a result of the atomic bomb testing.
>
> (https://collections.lib.utah.edu/details?id=1248
> 578&facet_setname_s=uum_dua)

The events and catastrophic process are perhaps less repetition than sedimentation: the toxin is embedded inside for generations to come.

In Kate Brown's *A Manual for Survival: A Chernobyl Guide for the Future*, she records:

> A commission arrived in April to take measurements. A brigade of doctors filed in to examine villagers. They found a lot of illness in the after-accident period, a doubling of the rate of birth defects, and a strikingly high incidence of infant mortality.
>
> (unpaginated)

The toxins had seeped into the very environment:

> Marei's team found that the swampy, sandy soils of the Pripyat Marshes were the most conducive of any soil type for transmitting radioactive isotopes into the food chain. Swamps in conditions of continual resaturation accumulate peaty soils that are rich in organic substances but poor in minerals. Plants searching for potassium, iodine, calcium, and sodium readily take up radioactive strontium, cesium, and iodine that mimic these minerals. Marei found that the indigenous berries, mushrooms, and herbs of the marshes showed a very high transfer coefficient of radioactive nuclides from

soils to plants. His team also discovered that seasonal floods spread radioactive contaminants "in a mosaic pattern" to places where floodwaters surged. As the boggy soils delivered radionuclides to plants, grazing farm animals magnified radioactive elements in the milk they produced. For Marei, the pathway was clear: water, soil, plants, animals, milk, humans.

(unpaginated)

It is this sense of permanent crisis and infinite temporality that constitutes the catastrophic sublime, and brings it in alignment with the ' "endlessness" which is then equated to the apprehension of infinity at work in the sublime feeling' (Cunningham 558). Multiple forms of depicting a scrambled but iterative temporality of the atomic/nuclear disaster may be found in the narratives, which contributes to the catastrophic sublime's evocation of a state of permanent/continuing crisis that is often beyond comprehension.

The constant emphasis on the endlessness of radioactivity, suffering and loss which we see in the nuclear narratives is remarkably akin to the apocalyptic sublime where, as Gunn and Beard tell us following the work of Frank Kermode, we are in a state of 'immanent apocalypse', or, as Kate Brown puts it, 'an acceleration on a time line of destruction' (unpaginated). In such an apocalypse, we are dwelling in the end period (Kermode 272). In other words, in an immanent sublime, we do not look forward to a time when the world will end, rather we experience what Jean Baudrillard termed a 'metastasis'.

> there will be no end to anything, and all . . . endings will continue to unfold slowly, tediously, recurrently, in that hysteresis of everything which, like nails and hair, continues to grow after death . . . At bottom, all these things are already dead and, rather than have a happy or tragic resolution, a destiny, we shall have a thwarted end, a homeopathic end, an end distilled into all the various metastases of death.
>
> (Cited in Gunn and Beard 274)

The catastrophic sublime as an aesthetic mode presents us with a situation beyond time.

The Heroic and the Catastrophic Sublime

The life of Gen and his young brothers as they try to take care of their ailing mother constitutes the ten volumes of *Barefoot Gen*. After the first volume dealing with the bomb, the series is primarily about life after the bomb, the continuing horror, and the heroism, of the boys facing disease, poverty, hunger on an unimaginable scale. The heroic in the catastrophic sublime is the battle, on an everyday basis, to find food, employment and security. The catastrophic aftereffects have to be overcome, as individuals and as a community, and it is this element that recalls the older notions of the sublime.

There is an element in the aesthetics of the catastrophic sublime of nuclear bombing/disaster accounts that recalls the traditional sublime. If the sublime, as Neil Hertz and others have argued, is about overcoming something threatening, then the accounts – visual and verbal – of individuals battling extreme dangers of radiation in Chernobyl, Fukushima and others are resonant of the traditional sublime.[20]

In *Chernobyl*, Valery Legasov seek Gorbachev's 'permission to kill three men', by sending them into the exploded reactor to open up a stuck valve. Gorbachev grants permission with the statement, 'all victories inevitably come at a cost' (episode 2). In another episode (episode 3), the miners, about to descend into the reactor, ask Boris Scherbina whether, after the cleaning up and their own deaths due to radiation, they would be 'looked after'. Scherbina, to Legasov's shock, says, 'I don't know'. And yet the miners go in.

Grigoriy Medvedev cites A.M. Petroyants, Chairman of the USSR State Committee for Use of Atomic Energy, justifying the disaster with: 'science demands sacrifices' (131). The first cleaning workers at Chernobyl did indeed sacrifice themselves. Korneyev's photograph of the mass of radioactive substance (identified after analysis as a mixture of silicon dioxide, titanium, zirconium, magnesium and uranium) inside the reactor in Chernobyl, Shevchenko's aerial photographs of the smoking reactor, the photographs of workers cleaning the reactor, accounts and images of residents trying to rescue, rebuild and care for the injured in the midst of the devastation in Hiroshima and Nagasaki, the clean-up team inside Fukushima as drawn by Kazuto Tatsuta in *ICHI-F*, are examples of the catastrophic sublime demanding and enabling a certain heroism in the face of extreme suffering.

The heroic response to the obstacles and dangers of the sublime was, traditionally, through an act of rationalization and contemplation in what Tom Cochrane identifies as the 'heroic model' of overcoming the feeling of self-negation in the face of the sublime object:

> we enjoy our capacity to engage with the sublime object, providing an enhanced recognition of our powers. And since many recognised cases of the sublime require only that the subject perceive or contemplate the object.
>
> (Cochrane 135)

And further:

> While one's sense of fear could be directly sensitive to the various threatening features, the enjoyment of one's powers is at best third hand; a response to the awareness of overcoming one's fear.
>
> (136)

Cochrane is himself not sympathetic to the reflective-heroic response to the sublime, treating it as inadequate in the face of threatening sights. A parallel to this theme of the radiation-hero, and of the heroic itself, is also found in other texts.

The *hibakusha* cinema of the 1950s, writes Donald Richie, was less about the horror of the events in Hiroshima–Nagasaki than about the 'tenacity of the inhabitants, their industry and bravery' (unpaginated). Images of nurses and physicians toiling at make-shift first-aid stations (*Flash of Light*, 72–7), the seriously injured policeman Tokuo Fujita issuing certificates to the injured at a roadside (52), bomb victims being rushed to relief stations (47), residents carrying human remains (206–7), civil defense units at work in Nagasaki (216–7, 218–9) instantiate what Broderick identifies in the cinematic representations, but they also do more.[21] In films such as *Inseparable*, about Chernobyl, the focus is on the clean-up workers at the site. Darmon Richter notes that the first workers and fire-fighters 'had been given little or no briefing about the nature of the accident' (70). Svetlana Alexievich notes that 'the Soviet Union sent 800,000 regular conscripts and reservist clean-up workers to the disaster area. The average age of the drafted workers was thirty-three, while the conscripts were fresh out of school' (3).

Heroic actions have traditionally been associated with sublime grandeur (Duro 47). The heroic personal narrative of ascent and conquest too, as commentators note, are a part of the aesthetic of the sublime (White). The catastrophe in the representations, while mourned and resented (especially in later writings from survivors), is also depicted as the time to *do* something. That is, the catastrophe is not the end of something, but the beginning of something like an individual but mostly communitarian *affirmation*.[22]

Michihiko Hachiya writes of a woman with extensive burns: 'her burns were not caused by the *pika* but by fire as she tried to rescue members of her family from their burning house' (74). Alexievich writes of the clean-up squads that went into the Reactor first:

> I have heard people say that the behaviour of the firemen extinguishing the fire at the power station on the first night, and the behaviour of the clean-up workers later, resembled suicide. Collective suicide. The clean-up workers often did the job without protective clothing, unquestioningly heading into places where even the robots were malfunctioning. The truth about the high doses they were receiving was concealed from them, yet they were compliant, and later even delighted with the government certificates and medals awarded to them just before they died. Many did not survive that long. So what are they: heroes or suicides? . . . Reports on Chernobyl in the newspapers are thick with the language of war: 'nuclear', 'explosion', 'heroes'.

(29)

Accounts of the Chernobyl clean-up operations abound. In the words of one such worker:

> "Such heat . . . With the slightest increase of temperature, the bitumen immediately caught fire . . . If you stepped on it, you couldn't put one foot in front of the other; it tore off your boots . . . And the whole roof was littered with luminous, silvery pieces of debris of some kind. We kicked them aside. One moment they just seemed to lie there, the next moment they would catch fire . . . "
>
> (cited in Plokhy unpaginated)

Serhii Plokhy then comments:

> What Shavrei and Pryshchepa were kicking aside were pieces of graphite and radioactive fuel. These radioactive materials were irradiating everything around them, first and foremost the crew members, who had no instruments to measure the radiation or proper gear to protect themselves against it.
>
> (unpaginated)

Plokhy writes of other such first responders:

> All three engineers-turned-divers would die of radiation poisoning within weeks of their heroic action.
>
> (unpaginated)

In these representations, what strikes one is that the response to a sublime threat is *not* contemplation or processes of rationalization, but self-annihilating action. It could be argued that the catastrophic sublime is the decimation of the self, but the moot point is: what is the individual's or community's agency in the face of the sublime? The above accounts seem to indicate that the response to an awareness of catastrophic events, the community and the individual responds with decisive, even suicidal actions. The annihilation of the self which the sublime engenders is, in such cases, a willed annihilation.

Notes

1 Hersey was criticized for a 'moral deficiency' in recording the sights of Hiroshima's destruction. Mary McCarthy argued that he had 'minimize the atomic bombing by treating it like an earthquake or hurricane or some other natural disaster' (Lifton and Mitchell 89).

2 Of the photographs of the atomic bombing, Carole Gallagher identifies two types: those that capture the mushroom cloud on the horizon and those that capture the devastated cities and injured humans (42).

3 Later, Peter Goin in *Nuclear Landscapes* (1991) would capture the vastness of the landscapes with the derelict and abandoned remainders of the nuclear

plants and factories. The difference, of course, is, by the time the plants and factories were abandoned, the landscape had been toxified beyond retrieval.

4 Masco also notes:

> Witnesses to the first atomic blast later evoked the sublime to capture its meaning, in many cases mediating the physical pain and intellectual pleasure of their technoscientific achievement through a deployment of religious imagery.
>
> (4)

And:

> The beauty of nuclear weapons science in Los Alamos has always been one of its most dangerous elements, allowing an aestheticization of scientific knowledge to circumvent the political import of engineering weapons of mass destruction.
>
> (19)

5 See Paul White's brilliant reading of Charles Darwin's account of the Concepción earthquake and the aesthetics of the 'geological sublime' where White detects the multiple registers of emotions and scientific inquiry, even as he makes 'a sublime spectacle of geology through its picturesque imagery and apocalyptic allusion' (59).

6 Critics have argued that the photographs of Yamahata lack a certain empathy and composed in too professional a fashion, so that they will possess aesthetic value (see Marcon).

7 Critics have argued that several of the photographs that exhibit signs of radiation damage actually embody a double exposure: 'flashed by innocuous and toxic forms of light' (Pringle 135). The damaged photographs and even equipment (such as the camera of Vladimir Shevchenko, who took the first pictures of Chernobyl, which is even now so radioactive that it has remained buried since 1986) are 'material witnesses' of an intangible, invisible toxin, such as radiation (Schuppli).

8 Masco adds:

> one of the first U.S. civil defense projects of the Cold War was to make every U.S. city a target and every U.S. citizen a potential victim of nuclear attack. The FCDA circulated increasingly detailed maps of the likely targets of a Soviet nuclear attack through the 1950s, listing the cities in order of population and ranking them as potential targets.
>
> (258)

9 Eubanks notes how the later murals embody a 'polyphonic engagement with issues of suffering, aggression and culpability' as the Murakis painted the death of American PoWs, the skulls of the Japanese dead merging with corpses of the American dead and the evocation of a 'sense of moral reckoning' in documenting the American killed by the Japanese (1621).

10 The Marukis who were not themselves in Hiroshima on 6 August 1945, painted them after talking to the survivors and eyewitnesses.

11 Yukinori Okamura argues about the murals:

> The significance of the Hiroshima Panels as a social movement lies in the way that they visualized the sufferings of nuclear disaster at the time when they were concealed because of political pressure. The panels were

meant to be not only a reminder of the atrocities of a past war but also a means of resistance to violence happening at the time and in future.

(523)

12 One of the most poignant scenes in *Barefoot Gen* occurs in volume III where the atomic bomb victim, Seiji, ostracized and shut away by his own family (Gen is hired to care for him), decides that he must paint the dead bodies: 'I'm going to draw the suffering faces of every one of these people – turned into monsters and tossed away like old rags' (115). The entire episode is a meta-comment on the commitment to an aestheticized memorializing.

13 The identification of the dead in Hiroshima and Nagasaki was an impossible task. As Susan Southard writes, after interviews with dozens of *hibakushas*:

> It would take five years for the city of Nagasaki to accomplish the nearly impossible task of counting the number of dead and injured from the atomic bombing. Officials lacked accurate population figures from before the bombing because older adults and young children had been evacuated, soldiers had been conscripted, and there was a lack of documentation for the thousands of Koreans, Chinese, and other Asian workers brought to Japan against their will. Tens of thousands of people, too, had left or returned to the city after the attack.

(unpaginated)

14 More than in the photographs, it is in the Maruki murals that it is possible to see the atomic sublime shading into the atomic Gothic, as Peter Hales has argued, with the grotesque images of suffering that populate the murals.

15 I adapt here Celia Lury's work on the increasingly topological turn in contemporary culture where lists, indices and icons enable ordering and transformation (2012). Whether this topological turn instantiates a mathematical sublime too is a moot point.

16 Workers cleaning up Chernobyl often 'fought against dosimeters'. The regulations stipulated that the maximum radiation permissible was 25 roentgen / hour, and once this limit was reached that person had to leave Chernobyl immediately (Borovoi 45–7). In addition to this, the authorities refused to believe it when a person's dosimeter displayed 1000 R/h (52).

17 In the face of overwhelming numbers of Japanese exhibiting signs of radiation exposure, the American Occupation forces refused to acknowledge that the injuries and sickness were radiation related. The strict censorship of all radiation-related news and even scientific studies during the Occupation years ensured that the Japanese and the American people were kept in the dark about the nuclear aftermath. As Janet Farrell Brodie puts it:

> U.S. officials controlled information about radiation f dropped over Japan by censoring newspapers, by silencing, by limiting circulation of the earliest official medical reports, by fomenting deliberately reassuring publicity campaigns, and by outright lies and denial... American officials confiscated Japanese reports, me slides, medical photographs, and films.

(845. Also see Lindee and Southard)

18 Cousins later initiated a 'Moral Adoption' programme to generate funds for the hibakusha and the Japanese children.

19 Nakazawa also berates those Japanese who made profits from the ABCC, like the man who collects bodies for the ABCC to dissect (V: 205–10).

20 In a pithy summary of the natural sublime, Rob Wilson writes:

> the natural sublime begets the resistance to a force that must be sub-
> jectively overcome through mighty language and psychic stance ("by a
> conscious and violent tearing away from the relations of the same object
> to the will"), that is, by a subject-altered "representation" of that same
> vast object . . .

(414)

21 Peter Hales has noted that the history of atomic science and nuclear power
    was often portrayed the triumph of heroic science (10). Admittedly, the fig-
    uration of kamikaze pilots and soldiers as heroic has been a cultural text for
    and by the Japanese too.
22 Johanna Lindbladh writing about Chernobyl cinema argues that such
    depictions moved beyond the negative aspects of the disaster to a sense of
    'something of a positive force contributing to rebirth' (241).

5 Planetary Precarity and Anti-nuclear Cosmopolitanism

Exposed to the 1 March 1954 nuclear tests in the Bikini Atoll, the *Lucky Dragon's* Ōishi Matashichi, besides documenting his experience on the boat, also notes the diffusion of the toxins across Japan:

> By April . . . terror of radioactive fallout became a reality from Hokkaido to Kyushu . . . On April 7, dust samples taken in Aichi measured 52 counts of radiation per minute. The count was up to 1,004 in rain in Shizuoka, 4,000 in Tokyo, 21,127 in Osaka, 2,300 in Hiroshima. Snow in Niigata and Sapporo, registered strong radiation . . . In Kyoto, the first 0.3 millimeters averaged 86,000 counts per minute and in Osaka, 24,000 – stunningly high counts.
>
> (36)

When James George brings together lives, characters and events from all over the world in *Ocean Roads*, he underscores their irradiated connections: they are all to varying degrees, children, families and relatives of the bomb. As Troy, the son of Isaac Simeon, one of the Los Alamos scientists, thinks of the nuclearized history of his family:

> The only things he knows about his father is that he was born in the deserts of New Mexico and died on a Pacific atoll. An atoll where they built the airfield that launched the B-29s that dropped the atomic bombs on Japan to end World War Two. His father's blood was in that coral runway. And that bomb was in his stepfather's blood.
>
> (121)

Later, we are told that Simeon's other son, Caleb, is dying of acute lymphoblastic leukaemia – a feature of *hibakushas* in Japan. In fact, Caleb is advised that he needs radiation therapy. When the doctor explains what the therapy does – 'it uses high-energy x-rays to kill the cancer cells' – Caleb starts to smile at the irony, because it was radiation that had given him the cancer in the first place, and he now needs more radiation to fight it (269).

DOI: 10.4324/9781003254294-5

In *Barefoot Gen*, Nakazawa's Gen, eight years after the Hiroshima bombing, admits after reading a news account that there has been a 'jump in cases of leukaemia among A-bomb survivors':

> And it terrifies me to think that I could suddenly come down with radiation symptoms too . . . It makes me feel so helpless.
>
> (X: 165)

On the pages following this incident, Nakazawa draws the city's roadsides: buildings and electric posts with posters 'March against the Soviet H-bomb test: Support Pope Pius XII's Proposal to ban Nuclear Weapons' (167). Hiroshima's survivors and victims are responding to a global threat, in the form of the USSR's atomic bomb programme and, implicitly the arms race.

Spread across time (generations) and space, the nuclear effect is evenly distributed, as George, Matashichi and Nakazawa imply. Debjani Ganguly writes about James George's *Ocean Roads*:

> While the novel's present is 1989 in New Zealand, its historical frame is haunted by anniversaries of nuclear and chemical warfare – the testing of the atomic bomb in the deserts of New Mexico; the bombing of Pearl Harbor; Hiroshima and Nagasaki; the VJ Day celebrations of August, 1945; and the 1969 napalm gas attack in Vietnam. Each of these is captured through the photographic lens of one of its key characters, Etta, a Maori woman and a Pulitzer Prize winner whose filial bonds stretch across these global histories of irradiated violence.
>
> (437)

Ganguly's phrase, 'global histories of irradiated violence', emblematizes the nuclear humanities' principal concern: planetary precarity. In such circumstances, there are no 'isolated' nuclear incidents or accidents. As Elizabeth DeLoughrey, examining the 'myth' of isolates when it comes to nuclear tests, puts it:

> The lie of isolation has indeed been a dangerous game, to the Marshall Islanders especially, and beyond. Due to these thermo-nuclear weapons, the entire planet is permeated with militarized radiation . . . Radioactive elements produced by these weapons were spread through the atmosphere, deposited into water supplies and soils, absorbed by plants and thus into the bone tissue of humans all over the globe. The body of every human on the planet now contains strontium[90] a man-made byproduct of nuclear detonations . . . At very conservative estimates, these nuclear weapons tests have produced 400,000 cancer deaths worldwide.
>
> (179)

Others have linked nuclear fallout monitoring to global climate change monitoring. Paul Edwards, writing a year after Fukushima in the *Bulletin of the Atomic Scientists*, summarizes:

> Without nuclear weapons and fallout, we might know much less than we do about the atmosphere. Without climate models, we would not have understood the full extent of those weapons' power to annihilate not only human beings, but other species as well.
>
> (37)

In *Chernobyl*, Valery Legasov is asked by Mikhail Gorbachev, the President, when things would be safe again, Legasov does not miss a beat and responds, 'not within our lifetime'.

All the above commentators have signalled one key point: there are *no* safe places or safe times from nuclear disaster and all parts of the planet are linked in entangled histories of nuclear precarity. What we are looking at, then, is *multispecies* vulnerability and precarity.[1]

Planetary precarity is engendered by the *nuclearization of the planet*, where all life forms and the non-living are subject to risk, and even distant spaces and elements – soil, earth and water – carry toxins from Chernobyl, Fukushima and nuclear tests. The toxins from Chernobyl's Reactor 4, we now know, spread as far as the United Kingdom, and had reached Sweden in less than 24 hours. Such a precarity acknowledges the 'one world or none' stance of the scientists and philosophers – from Einstein to Oppenheimer – in the 1946 volume that is an initiating text of the anti-nuclear campaign, in an ironic way.

Planetary precarity produces the advocacy and activist side of an atomic internationalism – or 'one worldism' – that originated, as critics have commented, in the aftermath of the Second World War, albeit fuelled by a fear of communism and situated within the context of the Cold War:

> In the case of the atomic bomb, intellectuals' responses transcended national boundaries. Not only was the atomic bomb's very construction in the Manhattan Project a transnational venture, in the sense that it gathered scientists from several different nationalities, but the atomic war was also a danger which transcended borders, creating a transnational threat. European intellectuals followed the debate in the US, where scientists and military experts intensively discussed the perils of using the weapons, as well as the potential peaceful uses of atomic energy.
>
> (Andren 3–4)

At events such as the 1995 'International Symposium: 50th Anniversary of the Atomic Bombs' with Hiroshima's survivors alongside the downwinders of Nevada and the survivors of *Lucky Dragon*, the '*hibakusha* from all over the world' came together, writes Matashichi (77), gesturing at a planetary mobilization of anti-nuclear sentiment.

In this chapter, I turn to the discourses and representational strategies in the anti-nuke movement, through a reading of texts from different places, from Navajo nation art to Maori fiction to documentary films about India's uranium mining. The subjects are as varied as the countries and cultures of origin, and the choice and approach is itself a meta-comment on the cosmopolitan nature of the anti-nuclear movement and its objects of protest: uranium mining, nuclear waste disposal, the non-human. Transnational memories of the atomic bombing of Japan or nuclear disasters such as Chernobyl bring together different nuclear subjects and landscapes into a broader ambit of 'global memory' (the subtitle of Ray Zwigenberg's book on Hiroshima), and this too is something the anti-nuclear art work and discourses capture. Like global memories of Hiroshima and Nagasaki that drive anti-nuke movements, protests about nuclear waste repositories also show signs of a more-than-local awareness and concern. For instance, in their study of public testimonies on nuclear waste repositories in four American states (Wisconsin, Maine, North Carolina and Georgia), Michael E. Kraft and Bruce B. Clary observed this pattern:

> Sixty percent addressed the consequences of a repository for their state as a whole, 28 percent talked about difficulties other states and the nation faced in dealing with DOE, and 11 percent discussed the international implications of the nuclear waste question. Only 23 percent spoke exclusively in terms of local impacts. The only notable variation among groups is that representatives of industry and utilities and of Indian tribes exhibit a more pronounced local, as opposed to state-level, geographic focus.
>
> (Kraft and Clary 98)

There is, evidently, the intersection of the local with the global and the planetary in discussions around nuclear power. In *Barefoot Gen*, Gen, the alter ego of Nakazawa, swears to 'make art that travels around the world . . . break down the narrow-minded barriers that people call national borders!' (IX: 136). By responding to and documenting the horrors of Hiroshima in a Japanese medium – manga – that travels around the world, Nakazawa does exactly that, and thereby contributes to an anti-nuclear cosmopolitanism.

Thus, it is by necessity that the nature of anti-nuclear movements is transnational. Just as atomic testing, disasters or bombing cannot be confined to a national or geopolitical border (indeed, in several cases, the tests were conducted in places far away from the USA or Europe, such as Bikini Atoll and the French tests in the Pacific), neither can the anti-nuke campaign. Moreover, the mining and trade of uranium is global in scope and nature. As Astrid Mignon Kirchhof and Jan-Henrik Meyer have pointed out, 'the anti-nuclear movement that emerged during these two decades – roughly between the first protests against the nuclear power plant at Fessenheim in France in 1971 and Chernobyl 1986 – was engaged in substantial transnational exchange' (168). The materials and discourses of protest examined here therefore cut across nations and local campaigns for precisely this reason, because a 'methodological nationalism' (Beck) is inadequate to address a technology and outcomes that call the entire planet's existence into question.

Admittedly, the strategies and modes of protest vary, from the confrontational to the non-violent, embodied and situated protests and public opinion influencing.[2] Yet, despite the variation within the repertoires of protests, there are connections, exchanges and overlaps. In any case, any work on planetary precarity needs to account for the larger role of such campaigns too.[3] Protestors against nuclear testing or waste disposal in Australia's outback also speak in terms of planetary precarity, just as documentation, in the form of artwork, of high levels of radiation in the plants of Chernobyl's Exclusion Zone brings attention to bear on the 'timelessness' of radioactive contamination and Edward Burtynsky's painting of Elliott Lake, Canada's remnants of uranium mining points to the denudation of the land and Peter Goin's photographs of nuclear landscapes in Yucca Mountains show craters left over from nuclear testing.

In what follows, I examine the mobilization of anti-nuclear sentiments, discourses and texts. Some of these derive directly from campaigns (including anti-nuclear waste, anti-uranium mining, anti-reactor) while others, by foregrounding the risks and threats posed by nuclear power, are in alignment with anti-nuclear sentiments.

'One World or None': Towards Anti-nuclear Cosmopolitanism

In her essay on anti-nuclear documentaries, Lisa Lynch speaks of an eco-cosmopolitanism. Lynch writes:

> Without denying that global warming is indeed a globalized problem, both films [*A Hard Rain, Uranium: Is It a Country?*] suggest that green nuclear energy is not the solution. Rather, they depict the concept of green nukes as part of a strategy that allows the first world to ignore the conditions of nuclear fuel production while also allowing multinational corporations to repackage an industry with known environmental hazards as an environmental enterprise. Keenly aware of both place and planet, these films can serve as a template for imagining a new, internationally based movement that takes a clear-eyed view of the relation between body, environment, and finance in the production of nuclear artifacts. As the cleanup of Fukushima

continues and countries around the globe detect persistent traces of its radioactive outflow, that movement may indeed be in the making. (346–7)

Lynch's emphasis on planetary precarity resulting from uranium mining, its global distribution and the resultant production of nuclear power is effected through a reading of Bradbury's *A Hard Rain*, noting how Bradbury 'rehearses and extends the arguments of earlier US antireactor documentaries' (342). While Lynch's focus is on relatively recent advocacy films, the theme of global precarity and the need for a transnational/ cosmopolitan approach to the ending or control of nuclear energy may be traced back even further.

'I write this as a warning to the *world*', was the opening phrase of Wilfred Burchett's essay, 'Atomic Plague', written after his trip to Hiroshima three days after the bombing and published in London's *The Daily Express* on 5th September 1945 (emphasis added). The shift outward from Japan to the planet itself in Burchett's rhetoric may be the founding moments of an anti-nuclear cosmopolitanism.

A year after Burchett's essay an unusual volume appeared: *One World or None*, with short essays by Arthur Compton, Philip Morrison, Albert Einstein, Robert Oppenheimer, Neils Bohr, Harold Urey, Leo Szilard, Eugene Wigner, Hans Bethe, etc. The essays, like Burchett's, emphasized planetary precarity and urged the nations to come together to control nuclear energy. (The control of nuclear energy and the need to create international systems to achieve this is an old suggestion, right from Neils Bohr's letter of the 9 June 1950 to the UN. To anti-nuclear activists, the problem lay with the hierarchization of science and industry where scientists and industrialists with the active cooperation of governments would control nuclear power. See Weart, Chapter 19.)

In his 'If the Bomb gets Out of Hand', an essay in *One World or None*, Philip Morrison, who worked on the atomic bomb and visited Hiroshima to study the effects of the bombing, speculates on the possible effects if the atomic bomb were dropped on *Manhattan*. First, Morrison describes the atomic bomb as a weapon of 'saturation': 'The atomic bomb is pre-eminently the weapon of saturation. It destroys so large an area so completely and so suddenly that the defense is overwhelmed' (Masters and Way 2). Then he begins his projection of an atomic war on America:

The streets and the buildings of Hiroshima are unfamiliar to Americans. Even from pictures of the damage realization is abstract and remote. A clearer and truer understanding can be gained from thinking of the bomb as falling on a city, among buildings and people, which Americans know well. The diversity of awful experience which I saw at Hiroshima, and which I was told about by its citizens, I shall project on an American target. Please do not believe that there is exaggeration here; this story will be conservative, it will allow for no increase; in the effectiveness of the bomb. It will tell of only one where, if there is atomic war, twenty will fall. Your city, too, is a good target.

(Masters and Way 3)

Morrison concludes with:

If the bomb gets out of hand, if we do not learn to live together so that science will be our help and not our hurt, there is only one sure future. The cities of men on earth will perish.

(Masters and Way 6)

Morrison moves from Hiroshima to America, from 1945 to some unspecified time in the future, and this makes a case for global precarity through nuclear war. In a similar fashion, Oppenheimer in the same volume would state:

The vastly increased powers of destruction that atomic weapons give us have brought with them a profound change in the balance between national and international interests. The common interest of all in the prevention of atomic warfare would seem immensely to overshadow any purely national interest, whether of welfare or of security.

(Masters and Way 25)

E.U. Condon, briefly a member of the Los Alamos project, declares: 'We cannot seek national security in armament in a world possessed of atomic arms' (Masters and Way 41). Harold Urey writes:

A world war in which atomic weapons are used might very well weaken all of our countries and peoples to such an extent that they would not be able to survive in the future. And not only may our own culture be destroyed by these weapons of mass destruction, but all civilizations as they exist in the world really be retarded and weakened for centuries to come.

It all adds up to the most dangerous situation that humanity has ever faced in all history.

(Masters and Way 53)

Albert Einstein opens his essay, 'The Way Out', with:

The construction of the atom bomb has brought about the effect that all the people living in cities are threatened, everywhere and constantly, with sudden destruction. There is no doubt that this condition has to be abolished if man is to prove himself worthy, at least to ·some extent, of the self-chosen name of *homo sapiens.*

(Masters and Way 76)

While many of the authors in the volume speak of international *control* over atomic weapons, the one central theme across the volume's distinguished essayists is that the entire world and human civilization is at risk, *irrespective of which country owns the weapons or who deploys them first.* That is, anti-nuclear cosmopolitanism emerges as a two-stranded argument: the distribution of weapons and nuclear sciences across the world's geopolitical formations (principally, Russia) *and* the potential for ruination of the world. The risk, the threat and the decimation, the volume suggests, is everywhere. The 'one world' is a world facing a *common* threat. The concluding essay by the Federation of Atomic Scientists states:

> *The nations can have atomic energy, and much more. But they cannot have it in a world where war may come.*
>
> (Masters and Way 78, emphasis in original)

In the documentary film, *One World or None*, made with the assistance of the Federation of American Scientists as the credits indicate, the opening images show the globe rotating inside the letter O in the words 'One', 'World', and 'Or', which appear one after another, before the word 'None' appears. After the 'None' appears, a small bright speck moves swiftly down from the top of the screen and smashes into the 'O' and the globe inside it. Then, an explosion rips across the screen and then there is only a bright light; nothing else is visible. When the voice-over begins, we are first told of the transnational, and cosmopolitan, nature of the scientific discoveries and theories that enabled the creation of the atomic bomb: the German Einstein, the Dane Neils Bohr, the American Anderson, the New Zealander Rutherford, the Austrian Meitner, etc. The commentator says: 'no one scientist or nation is responsible'. On the screen, the major elements involved in the fission process appear, and in the backdrop, the various national flags appear as thumbnails. The voice-over speaks of the 'pooling' of knowledge. Then, the story shifts to the possibility of an American city such as New York and San Francisco being bombed. A flash of light, with the skull inside, appears. The skull is placed in the middle of concentric circles that are animated, flashing, implying the expansion of the range of destruction – reminiscent of the photographs of ruined Hiroshima and Nagasaki in *Flash of Light*, arranged in increasing distance from the hypocentre, and attempting to measure the scale of destruction as we move away/outward.[4] Atomic energy, it concludes, can be for the benefit of all, in all nations, provided it is free from the menace of war.

Neils Bohr's letter to the United Nations used the definition of the term 'cosmopolitan' when he wrote:

> the proper appreciation of the duties and responsibilities implied in world citizenship is in our time more necessary than ever before.
>
> (Bohr)

The responsibility, he said, was/ought to be, directed at an 'open' world:

> An open world where each nation can assert itself solely by the extent to which it can contribute to the common culture and is able to help others with experience and resources must be the goal to be put above everything else. Still, example in such respects can be effective only if isolation is abandoned and free discussion of cultural and social developments permitted across all boundaries.

Writing in the wake of the Bikini tests, the Einstein-Russell manifesto of 1955 (to which Max Born, Frederic Joliot-Curie, Linus Pauling and others also added their signatures) begins by calling themselves 'not as members of this or that nation, continent, or creed, but as human beings, members of the species Man'. It called for a halt to nuclear weapons production but also to war. It urged everyone to think in terms of universals, including universal destruction:

> There lies before us, if we choose, continual progress in happiness, knowledge, and wisdom. Shall we, instead, choose death, because we cannot forget our quarrels? We appeal as human beings to human beings: Remember your humanity, and forget the rest. If you can do so, the way lies open to a new Paradise; if you cannot, there lies before you the risk of universal death.

Thus, in the texts of the first decade (1945–55), the emphasis was on the international control over nuclear power. They do not advocate ending the research into or developing nuclear power – just a mechanism to control its use. And this control, the text and film suggests, is the task of the world itself. All of this implies an anti-nuclear, global and *cosmopolitan* responsibility to save the world.

In celebrated documentaries such as *If You Love This Planet* (1982), Helen Caldicott, the anti-nuke activist who is also the protagonist of the film, goes into great detail about the dangers of nuclearization. But more importantly, for our purposes, she refuses to localize nuclear harm. Drawing attention to Hiroshima, she says: 'in Hiroshima, there was an outside world to come to help. There will be nobody [in the case of an all-out nuclear war]'. She then begins an inventory of mass deaths: how many Americans will die, then adds, also Canadian, Russian, British, European, most of the Chinese . . . Caldicott signals planetary precarity and global extermination. The pointlessness of assuming a limited war, localized threat and circumscribed destruction is highlighted in all its absurdity when Caldicott declares: 'this [nuclear war] is not a war. This is extermination'. Decades later, Ira Helfand, of the International Physicians for the Prevention of Nuclear War in *The Beginning of the End of Nuclear Weapons* (2019) would echo the same point as Caldicott:

> Even a much more limited nuclear war, one which involves other nuclear powers like India and Pakistan would in fact be a threat to the entire world.

With more rhetorical flourish, Arundhati Roy writes of the 1998 Pokhran nuclear tests, India:

> If only nuclear war was the kind of war in which countries battle countries, and men battle men. But it isn't. If there is a nuclear war, our foes will not be China or America or even each other. Our foe will be the earth herself. Our cities and forests, our fields and villages will burn for days. Rivers will turn to poison. The air will become fire. The wind will spread the flames. When everything there is to burn has burned and the fires die, smoke will rise and shut out the sun. The earth will be enveloped in darkness. There will be no day – only interminable night . . . The nuclear bomb is the most anti-democratic, anti-national, anti-human, outright evil thing that man has ever made.
>
> (6–7)

Roy's rhetoric, and dramatization, resonate with other anti-nuclear campaigns and their discourses.

Now, the dramatization of the possible destruction of the cities of New York or San Francisco constitutes the 'fabulously textual' (as Jacques Derrida called it, 23) style and structures the prognosis of a nuclear future. Alternating with the archival footage of the Hiroshima destruction, the animation of buildings falling and burning in films like *One World or None* erodes the documentary realism of anti-nuclear activism. The fabulously textual mode takes recourse to a figural realism – whether this is in the animated projection/fantasy of a future war in the films or metaphor of 'way . . . to a new Paradise' in the Einstein-Russell manifesto. In *A Hard Rain*, the effects of radon gas – one of the materials that emerges in uranium mining – are presented in interesting ways. We are first told how it affects minds, families and communities – and for the latter, the film shows Aboriginal families walking through the open land – before showing us a sleeping infant. As the voice-over (from Dr Rosalie Bertell, an epidemiologist) tells us – the invisible, odourless gas gets into the body, the film shows ghostly, wispy cloud-like vapours floating around the baby. Then, it shows an animation of lungs inhaling the gas, causing 'teratogenic damage'. In this case, the film moves inward – as opposed to the catastrophic realism which moved away from the hypocentre of the bomb farther and farther out. First the landscape, then communities and families, then the sleeping infant and finally the lungs. Also shifting the register from the aesthetics of the prospect view – the uranium mining area, the landscape – to the intimately textual of the infant accompanied by the ghostly) and finally the uncanny images in which the lungs appear as animation. Bradbury, like Bohr and *One World Or None*, seeks to emphasize the future-catastrophic by moving away from the realist mode of the documentary. So would such metaphoric and non-realist representation capture the frightening potential and risks of nuclearization?

As Hayden White argues, the debate is about 'the truth-value of a text which promises in its preface that "none of the facts has been invented" but whose meaning resides in large measure in the extent to which it copies the plot-structure of a poetic fiction' (117). Here the emphasis is less on the catastrophic realism with which Hiroshima or Nagasaki's destruction in the past is presented (the archival material, such as photographs), than on an infusion of anxiety, anger, hope and other strong emotions into images of a *future* destruction of the world itself. In White's terms, when writing of Primo Levi's witness account of the concentration camp:

> The most vivid scenes of the horrors of life in the camps produced by Levi consist less of the delineation of 'facts' as conventionally conceived than of the sequences of figures he creates by which to endow the facts with passion, his own feelings about and the value he therefore attaches to them.
>
> (119)

If the future can only be perceived as an absolute monstrosity, then the figuration of the monstrosity of destruction – or hope – demands a rhetoric such as the ones we see in the above examples of anti-nuclear films and texts. The figural realism, I suggest, is a political choice as an aesthetic because it rejects the necessity of local realities – the specific of cities, people, etc. – in favour of a *figure* of the human form or cities that *could* be destroyed. This aesthetic strategy, I suggest, is itself a universalizing gesture that signals a cosmopolitanism of representing the future destruction of the world.

The (Under)Mining of Land

In *Uranium: Is It a Country?* (2008), Reg Dodd begins by telling us that in an area of about 150 kilometres, there are 3–4 'big mines'. He informs us of the Aboriginals – he belongs to the Arabana peoples – who do not go near such mining places because they believe these are 'poisonous'. In the same film, Dave Sweeney of the Australian Conservation Foundation suggests that given the role uranium plays in war, the best thing to do for the world is to 'keep the genie in the bottle'. In Dodd, the discourse is of the toxification of the land, and in Sweeney, it is what the land is made to exude, spew that toxifies the *world* at large. This set of shots is interspersed with (i) the head of the uranium mining corporation who speaks in terms of the profits generated annually from Australia's export of uranium (Australia holds 30–40% of the world's uranium deposits) and (ii) an anti-uranium protest march.

Multiple discourses intersect here, from capitalism and its ruthless pursuit of profit to the local communities' ignorance, beliefs and fear around mining, the role of place/s and, finally, the 'saving-the-world'.

In Chantal Spitz's novel about nuclear tests in Tahiti, *Island of Shattered Dreams*, a character ponders over the impossibility of their (natives') situation with the arrival of the new technology and its accompanying horrors:

> How can anyone explain the inexplicable, when there are no words for it in their language: base – launch – missile – laboratory – experiment – nuclear? How can anyone pass on images that their minds cannot imagine?
>
> (79)

Spitz's character is addressing the key problem of comprehension: the Aboriginal/native dwellers who did not know or understand nuclear colonialism.

Three domains figure prominently in anti-nuclear discourses: uranium mining, nuclear testing and nuclear waste disposal, and in all these domains, the place is the focus.[5]

Place(ing) Protest

In the protracted Jabiluka anti-mining campaign, the emphasis was dual: the present act of mining and the consequences for the future.

In *Jabiluka* (1997), emphasis is laid on the fact that the land is the 'traditional home of the Mirrar people, and the world's oldest living culture'. Speaking about the land, Yvonne Margarula says in the film: 'we own the country, our country . . . black country, not white country'. Into this land, which has 'sacred sites of significance [to the Aboriginals]', comes the white man who, says the voice-over, 'understands the value of uranium in an energy-hungry world'. The mining, as the protesters and campaigners point out in films like *Jabiluka* and *Dirt Cheap 30 Years On: The Story of Uranium Mining in Kakadu* (2006), disturbed the traditional environment of the Mirrar peoples. In *Buddha Weeps in Jadugoda* (1999), a documentary on uranium mining in the Jadugoda region of Jharkhand state of India, the film opens with the origin story of the people of the place: the Santhals, Mundas and other Adivasis. The film focuses on their sacred groves and their rituals. A journalist from the community says in the film, the land's resources are 'not for our [tribals'] prosperity . . . they have become our curse'. The opening statement in the voice-over by traditional landowner Enice Marsh of the Adnyamathanha people, in *Nuclear Wasteland* (2016), a documentary about the Flinders mountain range chosen by the government as Australia's first nuclear waste dump, is 'to me it feels like a death penalty'. For the Adnyamathanha people, Flinders is 'sacred' because there are ancestral burial sites in the area.

Just like in the above films focusing on nuclear colonialism, the natives in Chantal Spitz's *Island of Shattered Dreams* eventually realize that the creation of a nuclear test site means the white man appropriates their sacred lands:

> On what Land will we bury our children's *pito*? The Land does not belong to us, it belongs to the generations to come. This is our home and we don't want their centre. Why don't they set it up in their own country?
>
> (79)

The residents are expelled, and they lose their homes.

This is the discourse of place that operates in the context of the anti-mining protests. The discourse has several components.

In anti-nuclear documentaries from countries as diverse as Australia and India, uranium mining is pitted as the binary opposite of, indeed as antagonistic to, cultural heritage and the values of the local populace and residents. The Aboriginals and *Adivasis* (forest dwellers of the Jadugoda region of India in the film, *Buddha Weeps in Jadugoda*) represent one specific set of values. The modern man (white or brown) represents the exact opposite.

There is the discourse of Nature versus culture, where the Aboriginal way of life is, not unproblematically, but perhaps as an instance of a strategic essentialism, valorized for being closer to Nature and less exploitative of it, while the modern white man's culture is brutal and exploitative in the extreme. Kakadu, Jabiluka and Jadugoda represent, in this rhetoric, places older than time, and the very antithesis of the contemporary. The films scan the landscape, the everyday lives of the residents, their rituals and songs and then cut to the containers, machinery, offices and the shipments of the mined products or the mining corporations. In the process, the anti-nuke films present us with two different ways of life: the Aboriginal hamlet *versus* Melbourne, for example).

In a mural such as 'Wheat-pasted pump house on a radioactive wasteland' for the Painted Desert Project in Cameron (https://hypera llergic.com/401017/navajo-nation-artists-respond-to-the-threat-of-uranium-radiation/), a wasteland 'produces' both, a socially invaluable instrument – a pump house – and an artifact. The pump house is first, the *antithesis* of the radioactive wasteland, because it stands for the possibility of irrigation and drinking water, and therefore life. Yet, the pump house is also an indexical sign of the nature of the natural resource beneath the ground: is the pump house pumping safe or toxic water for the community?

Now, the location of the protest – Kadaku, Jadugoda, Utah – is integral to the protest's rhetoric and dynamics. As Danielle Endres and Samantha Senda-Cook in their analysis of the role of place in social movements put it 'place functions along with other rhetorical performances in social movement discourse' (258). In the above texts around uranium mining, we see that 'place-based arguments discursively invoke images or memories of a place to support an argument' (258).

The Imperial War Museums' poster collection includes McLaren's *No* for Campaign for Nuclear Disarmament [CND]. McLaren focused not on the destructive power of nuclear weapons per se. Using the mushroom cloud as the entire top half of the poster enabled him to depict it as the roof of the diegetic space. Underneath it, nested almost like a home, McLaren draws the nuclear symbol. Then, he lists the costs of each item of war and the economic costs. What would the price of a V-bomber or a Polaris missile be, and what *could* that amount of money do for the society.

In *Buddha Weeps in Jadugoda*, we are shown scenes of rituals and ceremonies being held in the community. There are also scenes of community prayers, feasts and cultural festivals. In these scenes, like in the shots from other documentaries which show us the indigenous populations in their everyday lives, the place functions as material rhetoric. Endres and Senda-Cook write:

> material rhetoric not only focuses on material structures but also the symbols that are interrelated with these structures. Many protest events encompass this fluidity between the material and the discursive because they are held in places with symbolic meaning or are meant to alter or challenge the dominant meaning of a place. While we consider how material structures are rhetorical, in part, because of their symbolicity, we also examine how these physical structures have material consequences.
>
> (262)

The emphasis on local cultural practices, memories and values of sacralized places in anti-mining campaigns renders the place both material and symbol, and constitutive of the very rhetoric of protest. The protection of the sacred places, as we seen in all these protest rhetoric, is offered as a counter to the nuclear colonialism (and its rhetorical moves, including the rhetoric of nuclear stewardship, in the post-Cold War period) that has been disrespectful of the place.[6]

There is another aspect of the material rhetoric of place. The battle is against the *local* source of materials that fuel massively catastrophic wars

Figure 5.1 Ian McLaren Poster for CND, 1965.

Source: Reproduced with the permission of Professor McLaren and the Imperial War Museum. Ian McLaren Poster for CND, 1965 (www.iwm.org.uk/history/six-protest-post ers-from-the-1960s-and-1970s.)

which are often fought elsewhere. The protests therefore invokes places from elsewhere as constitutive of protests *here*. As the protestors' placards declare in *Uranium: Is It a Country?:* 'stop uranium mining to stop wars'. That is, it is not war *per se* that is being protested, but the backdrop to the manufacture of more and more dangerous weapons, through uranium production and the transnational impact of, say, Australian uranium production on wars in other parts of the world. This is a transnationalization of the material rhetoric of those places that are marked by the presence and extraction of uranium. The other parts of the world are material embodiments of Australian uranium, so to speak. The uranium in an Australian mine is transnationalized, just as a war with nuclear weapons is transnational and planetary in scope. This is the material rhetoric of uranium mining and production.

Protestors at sites such as Jabiluka or Jadugoda or Utah place their bodies *at the site*, thereby transforming the places into 'embodied rhetorical performances' (Endres and Senda-Cook 263). Scenes of the protestors lying down, pushed, shoved and beaten, and often arrested at various sites – from cities to the mining areas – achieve multiple effects.

First, exposing the body in the course of the protest to police reprisals, argues Judith Butler, serves a political purpose (2015). The bodies – which are of locals *and* protestors from outside – braving reprisals reaffirm the embodied nature of place. The places are those from which bodies have been displaced to enable uranium mining or nuclear tests, as in Maralinga or the Marshall Islands, or whose presence has been ignored as in Utah and Jadugoda and Jabiluka. In Butler's words:

> When the bodies of those deemed "disposable" or "ungrievable" assemble in public view . . . they are saying, "we have not slipped quietly into the shadows of public life: we have not become the glaring absence that structures your public life."
>
> (152)

Then, the place endows the diverse bodies – locals, journalists covering the event, participating outsiders – with material and symbolic value: not only as vulnerable bodies (because the materiality of the bodies can be damaged) but as bodies adhering to the values of *that* place.

Aboriginal protestors like Yvonne Margarula, who *belong* to the land where uranium is being mined, were deemed as *trespassers* by the state (the Court ruled against her petition that she cannot be treated as a trespasser). Thus, when Margarula and her associates from the community come to occupy the traditional, sacralized land in an attempt to demonstrate that they belong there (and not the white man with the mining machinery), the place – Jabiluka itself – rather than the individual becomes the 'face' of the struggle. I adapt here the work of W.J.T. Mitchell on the 'Occupy' movement. Mitchell proposes that the Occupy campaigns lacked a 'definite form or figure other than the dialectical poles of the mass and the individual, the assembled crowd and the lone, anonymous figure of resistance' (9). Anonymous individuals and the masses of protestors perform, in Mitchell's terms, an 'occupatio' which he defines as, 'taking the initiative in a space where one knows in advance that there will be resistance and counterarguments' (10). Expanding on Mitchell's argument, one could argue that the Aboriginals and indigenous peoples who come ('trespass') on the sites of mining or testing are in fact the *prior* occupants of the land or *pre-occupatio*. The *OED* tells us that 'preoccupied' is 'to occupy or take possession of (a place, etc.) in advance or before another person', and when employed in this sense, as I do here, it underscores prior rights – that of the Aboriginals, rights which the white man and the mining corporation have taken away. I also employ 'pre-occupatio' to signify 'being preoccupied with': in this case preoccupied with the loss of the sacralized and symbolically significant place.[7]

As noted earlier, the symbolic valence of the land for the indigenous people, the pre-occupatio of the place, is immeasurable. One of the primary contributors to this valence is the antiquity of the natural landscape *and* the antiquity of the residents of the place ('our people have been here for hundreds and hundreds and hundreds of generations', says Enice Marsh in *Nuclear Wasteland*). An increased emphasis on the antiquity of the place and its residents – pre-occupatio – therefore marks anti-uranium mining rhetoric. This emphasis on time makes interesting shifts. The Aboriginals are represented as predating the white man's culture, and as old as the earth. The timelessness, effectively, of the Aboriginals is treated as a precious and invaluable feature because it presents the origins of mankind itself. This timelessness (unfortunately) is also romanticized by scenes that set the Aboriginals in their 'natural' environs, whether it is the landscape or their hamlets and homes. The cut to the modern white man, his machinery and cities, shiny and organized in entirely different fashion automatically signals an evolutionary scale: where the Aboriginal is the manifestation of the earlier stage of humanity, the city and uranium mining with its elaborate machinery and processes, represents the later stage. And this is where the narratives become complicated. If timelessness is valued in and for itself, nuclear power and its consequences proffer the most horrific timelines imaginable: the half-lives of the uranium-related elements are measurable in tens of thousands of years. When the Jabiluka campaign poster declares 'reclaim the future', it is the dual-edged nature of time that is being invoked: there is *no* future for humankind unless we stop all nuclear activities whose timelines are beyond human lifespans. If the Aboriginals represent the timeless antiquity of mankind, uranium and its by-products represent the timeless futurity of toxicity. Anti-uranium mining discourse operates with this polychronic state of both, civilization and toxicity.

The *Nuclear* terra nullius

Nuclearization dismantles places that have for centuries supported indigenous populations. Hence, in most cases within these discourses, there is an overwhelming emphasis on the effects of 'nuclear colonialism' (Endres 2009) on indigenous peoples.[8]

Before turning to nuclear colonialism and its effects on the indigenous, it is important to note that this modality of colonial domination of the natives in places where natural resources that European capitalism and modernity desired is an extension of older colonial practices as well. Gabrielle Hecht has observed:

> The history of uranium mining, for example, shows that colonial practices and structures were appropriated – *not* overthrown – by the nuclear age, and proved central to its technopolitical success. Hiroshima uranium came from the Belgian Congo. After the war, Britain's colonial ties to uranium-supplying regions in Africa and Australia helped maintain nuclear relations with the US. South Africa's eagerness to place its vast uranium reserves at the disposal of the West led the US and Britain to gloss over the emerging apartheid regime. France could pursue an independent nuclear program because it had access to uranium not just on metropolitan soil, but also in its African colonies. Internal colonialism figured too: in the US, the richest uranium regions proved to be on Native American lands on the Colorado Plateau, while Australians found much of their uranium on Aboriginal lands in the Northern Territory. I could continue to enumerate examples: the Soviet Union mined uranium in East Germany and Czechoslovakia; South Africa mined uranium in (present-day) Namibia; Canada on native lands, India on tribal lands. And on and on. And the same for nuclear testing, as the US tested its weapons on the Marshall islands, France in Algeria and Polynesia, and Britain on Aboriginal lands.
>
> (3, emphasis in original)

Even nuclear alarmism in the rhetoric of Euro-Americana carries vestiges of colonial discourse: in this case, it is constantly reiterated that nuclear weapons are better off in the hands of Europe and America rather than in the hands of Third World nations (see Gusterson 1999, Biswas 2014). Thus, uranium mining and its deleterious effects born of white erasure of the indigenous has a long history – and this is the subject of anti-nuclear campaigns since the post-Cold War period.

In *Uranium: Is It a Country?* Michael Angwin of the Australian Uranium Association speaks of the dollars earned through uranium mining and export. The emphasis on profits and national interest that this discourse operates on implicitly calls for sacrifices on the part of those communities affected by the mining operations (Enders 45). The CEO, Chamber of Commerce and Industry, South Australia, Nigel McBride, in *Nuclear Wasteland*, admits that there are about one hundred nuclear waste sites across the country. Then comes the horrific statement, 'probably not stored particularly well . . . really not an ideal position'.

In sharp contrast to the CEOs in these films, the community leaders speak of the loss of their lands, their rights and their memories of the place. Thus, in the case of nuclearization, the creation of mining areas, quarries and factories render the place of indigenous occupation and residence into non-places, in the sense Marc Augé uses the term. In the process, the indigenous people become non-persons on the sites: they are replaced by engineers and technicians. In effect, the non-place is constructed through the operationalizing of an older discourse: of the *terra nullius*.[9]

As a component of natural law, Andrew Fitzmaurice has persuasively argued, the *terra nullius* idea shaped the mentality of Europe (2007). It was instrumental in how lands were organized for European consumption in Australia (Borch 2001) and North America (Buchan 2007. Also Allen 2003). But the postcolonial state, it appears, has adopted the discourse of *terra nullius* for its own purposes in countries like Ethiopia (Makki 2013). Or, as Uditi Sen (2017) in her study of the Andaman Islands has demonstrated, the Indian state's operations in the region carry a *de facto* logic of *terra nullius*, and so it must be treated as an instance of settler colonialism. Thus, when the Australian or Indian state installs mining and associated operations, it becomes an expression and extension of the state's sovereignty over the land and its people.

In every protest campaign, the indigenous peoples speak of their displacement from the land of their ancestors. The white man and his machines have settled into their lands, and desacralized it for good through two modes. First, the intrusion of men and materials that arrive as material embodiments of a distinct culture and process – extraction of natural resources. Second, the resources from under the ground are actually sold in international markets and to foreign nations. That is, the sacred land with its special significance has been treated as not just *terra nullius* – devoid of people, or nobody's land – but as being on par with any land with some resources as valued by the white man. Further, by transporting the materials from under the soil to different parts of the world, nuclear policy has redistributed the sacred land, fragmented it. The objections of the indigenous people who do not wish to be incorporated into the 'new' landscape of mining and extraction are overruled – exactly as settler colonialism operated a century and half earlier in Australia or Canada.

This is the nuclear *terra nullius*: the operations of the state and the corporation(s) empty the land of its rightful owners and residents but, more importantly, render the region and the land unfit for the future as well. That is, the nuclear *terra nullius* is simultaneously about the past and the future. The land, the dwelling place of the indigenous people for centuries, has been deemed empty of people and full of resources. In other words, nuclear colonialism proceeds with a double movement: of emptying the land of its residents and treating it as 'full'. The uranium beneath the soil makes the land rich, and the people living on this land are of no account. The nuclear *terra nullius* therefore makes the people exterior and extraneous to the *matter* underneath their feet. The fact that this same land carries the memories and bones of the residents' ancestors is, likewise, dismissed with the focus exclusively on the minerals. The people are excess, surplus, but the minerals are not.

It is in the same double movement that the land has been rendered toxic – inducing displacement and evacuation of the residents – for the inhabitants for an interminable future. With the mining of uranium, and the toxic wastes, tailings and outflows, the land's 'use value' has been irreversibly altered for the Aboriginal landowners. The toxification of the land is not, however, only from the mining: the very sacrilegious presence of the outsider on holy lands, combined with the extrusion of the minerals from beneath the earth, together render the land toxic.

The *de facto* logic of the *terra nullius* is operationalized, and made visible, in a different mode in the case of nuclear testing and waste disposal sites – and this is documented as part of the art of anti-nuclear campaigns.

Peter Goin captures the nuclear *terra nullius* in the photographs in the volume *Nuclear Landscapes* (1991). He sets up the sequence of photographs through a photographic textual framing, as Edmund Lee and Shirley Ho term it. He begins with photographs of the Trinity explosion, the construction of the Hanford Nuclear Reservations, the Marshall Islands, Nevada tests and the landscape at the test sites. Goin's introductory section ends with:

> The Trinity Site is now a National Historic Landmark . . . Bikini and Enewetak atolls are supposedly cleaned-up, and only the bunkers remain. More than 30 years have passed since the last nuclear detonation at Bikini and Enewetak, and the islands are becoming overgrown with vegetation. Many of Hanford reactors have been decommissioned.
> Yet the sites are radioactive.
>
> (27)

Thus, Goin has already indicated the nature of the landscapes we will be seeing: they are radioactive. Yucca Flat now shows signs of life, with Joshua trees on the desert floor in one photograph (39).[10] Remnants of buildings set up for various purposes in these areas dot the land, and some apparatuses, such as an army tank in Yucca Flat, 'remains highly radioactive' (40–1). Elsewhere, in the South Silent Canyon of the Nevada Test Site, the canyon's neighbouring cliffs show signs of accelerated erosion (52–3), the fault lines have shifted on the valley floor of Yucca Flat (54–5) and subsidence craters appear in other cases (50–1, 58–9, 62–3). Goin in these pictures focuses on the geological damage from nuclear tests. (The Yucca Mountain was authorized as America's first high-level nuclear waste repository in 2002, before federal funding was stopped in 2011.)

In 'Orchard Site', Goin examines the Hanford site. These were originally 'successful farming regions', and hence hardly *terra nullius*.

The 'orchard was cut down after the area was condemned for use in the Hanford Nuclear Reservation', he informs us (66). The photograph shows a landscape of scrubs with hundreds of tree stumps (67). It is exactly what it looks like: a site of deforestation. Except that the deforestation is nuclear-driven. Goin also photographs cement debris – which is radioactive – at Hanford (70–1). The waste storage pond has water which is 'contaminated with radioactivity' (74–5). There are landscapes with 'yellow posts [that] identify buried radioactive waste and potential surface contamination' (76–7).

Three photographs, all titled 'Burial Ground', from the Gable Mountain Pond waste burial site capture a landscape where there are 'massive storage tanks containing hundreds of millions of gallons of liquid radioactive waste'. Goin writes: 'some reports indicate that the tanks are already leaking' (82. Photographs: 83, 85, 87). The Retention Basin area has 'surface contamination' (90–1). The fact that Goin uses the term 'Burial Ground' is evocative: many of the lands across the world in which nuclear wastes are interred are also sacred lands for the indigenous peoples of the region.

Figure 5.2 Peter Goin, 'Burial Ground', Gable Mountain Pond.

Source: Reproduced with the permission of Professor Goin, from Nuclear Landscapes (Johns Hopkins University Press, 1991).

After covering the USA, Goin turns to the Bikini and Enewetak atolls. He photographs the massive 'crypt' storing '111,000 cubic yards of radioactively contaminated soil' (104–5). In 'Nuclear Bunker Complex', we see a patch of green adjacent to the concrete structure. Goin dispels any hope one may take from the lush green when he writes 'this area is still radioactive' (106–7). Debris scattered across the Islands are 'still radioactive' (116–7). 'Radioactive vegetation' covers the bunkers elsewhere on the Atoll (130–1). In his commentary on 'Bunker' (136–7), Goin summarizes the nuclear *terra nullius*:

> According to a Department of Energy survey report (April 1989), if islanders were to resume living here, they would be receiving a 20-millirem dose of radiation over and above normal background radiation. The surface soils of Enjebi Island are contaminated with residual levels of plutonium (half-life of 240, 360 years) and americium, which could be re-suspended by the wind and present an inhalation hazard.
>
> (136)

In short, the Island, if peopled, poses a threat that is likely to last a very long time, as the half-life span indicates. For his last photograph, 'Tide Pool', Goin writes: 'The Marshall Island sites of Bikini and Enewetak are still radioactive' (144).

Goin captures empty lands. However, as his comments indicate – and it is common knowledge too, now – the Bikini and Enewetak were inhabited spaces, as were lands in the Hanford area. These had been *emptied* and the natives relocated to facilitate the tests. This is the first iteration of the nuclear *terra nullius* and is deeply disturbing. The second iteration is even more frightening. Goin's landscapes are contaminated for a very long time. The emphasis on leakage of radioactive materials, the strewn radioactive debris and finally, the very nature of radioactivity which is unlikely to cease for the term of humanity's natural life – with half-lives of 240, 360 years – ensures that the area has to remain *terra nullius*, unless humans wish to absorb more than the acceptable limit of radiation if they return to the Atolls. In other words, it is a permanent *terra nullius*.

The derelict machinery, tanks, barbed wires and buildings are reminiscent of any other post-industrial ruin. Tim Edensor says of such ruins:

> The topographies of yesteryear thus reassert themselves in memory, the familiar crowd of industrial buildings and the fixtures and local amenities which supported them and their workforce, and in the ghostly traces of past embodied enactions, to produce a sort of phantom network.
>
> (148)

I have elsewhere argued about the Union Carbide plant in Bhopal, India: 'the post-disaster ruin which is UCIL is the effect of a *merged* metabolization, of nature (bodies, soil, air, water) and culture (chemicals from the plant)' (110, emphasis in original). The 'phantom network' is of course *material* in the sense the derelict material is still visible and tangible in all the test sites. In some cases, such as Bikini Atoll, they have merged with the environment (corals) too. What the Atoll or Nevada or Hanford symbolize in Goin's images is the abandonment of places after humankind has disarranged it forever. That is, *terra nullius* here is the ruin humankind leaves behind, tangible in the form of material ruins and invisible in the form of radioactivity. It is literally, nobody's land, because nobody *can* use it, live on it, cultivate it, without serious risk to their lives and that of their future generations. *Terra nullius* then is a material discursive act, a place or land where man has left a 'growing human footprint', in Dipesh Chakrabarty's evocative phrase, thereby denuding it of all potential (27).

I now turn to another aspect of the nuclear *terra nullius*: that of nuclear waste.

'Waste' in public discourse and understanding is associated with useless or extravagant consumption (Engler 60). It also signifies 'excess' and 'surplus' (Morrison 33). Debris and waste can function as an alternate site of reading history (Yaegar 106), or a 'catalogue of trauma' (105). But waste is also a 'product of time . . . an end product and the end of all living things . . . waste itself is a historical force; it becomes monuments to catastrophic loss' (Hawkins and Muecke xiv).

The point, and problem, with nuclear waste is: unless contained, the waste matter continues to radiate the neighbourhood. That is, the lands remain at risk from nuclear materials, even (especially?) if it is nuclear *waste*. In this process, the waste of the white man's industry and science – nuclear – reinscribes the *history* of their domination of the indigenous peoples. Or, shall we say, even the white man's mass produced waste has the agency to contaminate the indigenous peoples' lands and lives. The 'imposition' or 'dumping' of nuclear waste in places of indigenous residence is nothing short of genocide, argues the Indigenous Environmental Network (IEN) in its 1996 'Indigenous Anti-Nuclear Summit Declaration':

> The nuclear industry which has waged an undeclared war has poisoned our communities worldwide. For more that 50-years, the legacy of the nuclear chain, from exploration to waste has been proven, through documentation, to be genocidal and ethnocidal and a most deadly enemy of Indigenous Peoples.
>
> (Indigenous Environmental Network)

Waste of one race/nation is genocidal for a whole population of another, as the IEN declares:

The Indigenous Environmental Network opposes the recent decision of the United States President George W. Bush designating Yucca Mountain in Nevada as the country's official repository for highly radioactive nuclear waste. This is a wrong decision. Based upon scientific studies, Yucca Mountain is not a suitable site for a nuclear waste repository . . .

According to the spiritual leaders and tribal elders of the Indigenous tribes of *Western Shoshone* and *Paiute*, the Yucca Mountain is sacred with the regional area having deep cultural and historical value to their peoples. President W. Bush and many leaders of Congress do not respect these deep spiritual values and cultural life-ways that have sustained the Indigenous peoples of this region since time immemorial. In the eyes of Indigenous peoples that follow the traditional teachings of our tribal ways, this President and people in Congress do not have a heart of love and compassion for Life and have clouded minds that put money above the health and safety of people and all Life.

If the Yucca Mountain site is approved by Congress, it will store a total of 77,000 tons of highly radioactive waste, most of it spent fuel from nuclear power plants. The spent fuel, which will remain dangerous for hundreds of thousands of years, is now stored at dozens of power plant sites around the country.

If Congress allows the Yucca Mountain site to be approved, it would begin the largest nuclear waste transportation campaign in history, possibly endangering residents in 44 states, thousands of towns and cities, and tribal territories.

(Indigenous Environmental Network)

Terra nullius, then, is a history of erasure in the past but also in the contemporary. It is the erasure of people and their memories, since nuclear wastes make the land irretrievable, as the IEN statements indicate, for burials or collective memories. Although, as Yaegar and others have noted, waste is itself an archive, nuclear and radioactive waste is an archive that cannot be retrieved without extreme danger. The land is a material witness then, of invisible and dangerous erasures: like the tree stumps in Hanford, the craters in Yucca and other places pointing to something more insidious in the soil and water.

If Goin photographs a *terra nullius* where the poisoning is insidious and invisible, Edward Burtynsky paints the very source of the poison: uranium tailings in 'Uranium Tailings #12, Elliot Lake, Ontario, Canada' (1995). Tailings contain over a dozen radioactive nuclides and so, if left overground, can be dispersed through the air. Burtynsky, like Goin, shows the tailings merging with the land so that the contaminating element becomes naturalized. Elliot Lake, once a major uranium mining town before operations were shut down, has evidence of such a naturalization. The vegetation has been denuded and replaced, if that is the word, with sticks embedded in the tailings. The tailings have perhaps toxified the place so much that nothing grows. Burtynsky perhaps is moving beyond the terra nullius to *terra 'vita' nullius*: a land with any life, or without the possibility of any life. That is, it is not only human life that is rendered difficult (impossible?) as a result of the toxification of the soil and water: it is all life.

The older discourse of *terra nullius*, then, is a self-fulfilling prophecy, as the anti-nuke movement's art and rhetoric indicate, and must be expanded as a *terra vita nullius*: a land that does not allow/belong to any life form.

Exposed: The Nuclearized Non-human

Goin photographs a landscape of coconuts on Eneu Island, Bikini Atoll. However, these are unsafe. Goin says: 'the residual levels of cesium 137 (half-life of 30 years) absorbed into the coconuts . . . are still too high for human consumption to be safe' (124–5). He titles the image, 'Coconut Graveyard'. The last photograph, 'Tide Pool' shows electric cables intertwined with coral in the Bikini and Enewetak atolls. Goin tells us:

> even the best cleanup efforts could not remove all the electrical wire, pipe, and assorted smaller objects. Some were welded naturally into the coral in tidal areas throughout Bikini and Enewetak atolls, creating an unnatural beauty.
>
> (144)

What remains of the 'natural' in the collection of coconuts is the key question.

I have already referenced Burtynsky's painting that suggests denudation of local vegetation.

Attention, even inadvertent, to the non-human consequences of nuclear testing, atomic bombing and nuclear waste disposal has been steady in commentaries, the work of artists and writers. Broadly studied under the rubric of 'ecological' or 'environmental' consequences of nuclear power, the effects of radiation on plants, animals and geology have formed a crucial focus area of anti-nuclear movements around the world, since these, along with the human costs, constitute an index of planetary precarity.

In an early essay 'Effects of Nuclear Testing on Desert Vegetation' (1962), Lora Shields and Philip Wells, observing the vegetation in Yucca and French-man flats in the northern Mohave Desert, comment:

> It might be expected that after 8 years of nuclear testing very little vegetation would remain. On the contrary, except for the usual barren playa, no part of this basin lacks flowering plants. At a distance of 2 mi from most "ground zeros" the vegetation shows no

visible effects of weapon testing. Grotesque Joshua trees . . . relieve
the gray monotony of the dominant shrubs, hopsage . . . and
blackbrush . . .

(38)

They note that 'pioneer' species – non-native to the region – 'appear to
be invading at the perimeters of the totally denuded portions of certain
ground-zero areas' (40). This in itself does not strike the observers as odd:
that the native vegetation has entirely disappeared and that the so-called
return of nature is an effect of human engineering even when species-
change was not the intended effect of this engineering. In other words,
the replacement of species was a biocultural phenomenon induced by the
nuclear tests in the region.[11]

Paul Boyer cites an example of the media discourse around the non-
human effects of nuclear testing:

> In its August 11, 1951, issue, to cite only one of hundreds of
> examples, *Collier's* magazine published "Patty, the Atomic Pig."
> The article was based on an actual incident in which a piglet that
> was part of the "Noah's Ark" of goats, pigs, rats, and other experi-
> mental animals assembled for the July 1946 Operation Crossroads
> nuclear test at Bikini atoll was later found swimming in the radio-
> active waters of Bikini lagoon. The *Collier's* story . . . Illustrated with
> cute drawings, the story imagined Pattys thoughts before the blast . . .
> and her adventures afterwards . . . Patty not only survives (no scary
> radiation-exposure hazards here!) but grows to be a six-hundred-
> pound porker under the benevolent care of kindly scientists at the
> Naval Medical Research Institute at Bethesda, Maryland, and ends
> her days as a coddled exhibit in a zoo. Sugar-coated propaganda like
> this, part of a mountain of material in the media that reinforced the
> government's version both of the 1945 bombings and of Washington's
> subsequent nuclear program, served to deflect and neutralize serious
> scrutiny of the meaning and the implications of atomic weaponry,
> past, present, or future.
>
> (253–4)

Such reportage was meant to reassure the public about the nuclear power
being unleashed. In sharp contrast, for the locals, the visible signs of radiation sickness were present on the plants and animals of their region. In
the Cedar City area, for instance, one sheep farmer records:

> We started noticing white spots coming on [black sheep's] wool, on
> the top of their backs. We had a black horse there at camp, and pretty
> soon he had white spots come on his back and on his rump and all
> over, from this fallout, I guess.
>
> (cited in Fox 61)

Sarah Fox documents several such case histories of the consequences of nuclear testing on animals in the American west. Peter Goin photographed an enclosure once used to pen animals and study the impact of the nuclear explosions on them (47).

Photographs of pine trees (145 of them), specially planted in the Nevada test site to monitor the impact of nuclear explosions, showed them swaying hard under the impact. Elsewhere, in regions like Puerto Rico, experimentation with radiation was conducted by the USA, so that these regions became 'nuclear laboratories' (DeLoughrey 172). These places became, as DeLoughrey has argued, both laboratory and eco-system, where the latter was deemed to be a 'closed' one, and writes:

> Rethinking the ways in which science used the isolated island con-cept to produce some of the most apocalyptic technologies on earth challenges both the assumption of the primitive ahistorical island and what constitutes the laboratory itself.
>
> (175)

Indeed, echoing deLoughrey's argument is Grigoriy Medvedev in *Chernobyl Notebook*, who speaks of the ancient land of the area and the 'sense of the primordial' (26). Into this primordial area would arrive the laboratory, the manufacturing unit and radiation.

In the series *Chernobyl*, much of episode Four is concentrated on the killing of animals – cats, dogs, milch cattle – in the area because they had imbibed radiation and were potentially lethal. Rebecca Johnson in *Chernobyl's Wild Kingdom* (2015) documents the effects of the dis-aster on animals – years after the explosion. What she maps is a post-natural scene.

There was, contrary to expectations, a return of life forms in the Exclusion Zone:

> Beyond the fence, the vegetation is lush and green. In some places, trees grow right up to the edge of the road. Their branches arch overhead to form a leafy canopy so dense it filters out some of the sunlight and casts deep shadows all around. Clumps of grass and weeds sprout from cracks and potholes in the crumbling asphalt, and wildflowers bloom along the roadside.
>
> (18)

But it is not the life forms *per se* that Johnson is interested in: it is the nature of their nature, so to speak, their habitats and the food they con-sume. Commenting on the animal photographs by Sergey Gaschak, Deputy Director Science, Chornobyl Center for Nuclear Safety, she writes:

> more than fifteen hundred species of plants, mosses, and lichens are now growing in what amounts to ground zero of the nuclear

accident . . . Gaschak has spent a lot of time looking at wildlife . . . has seen moose with newborn calves crossing marshes, knee-deep in mud saturated with the remains of nuclear fallout. He has seen wild boars rooting for mushrooms around the gnarled roots of trees that have been growing in radioactive soil for years. He has even spotted birds nesting in the nooks and crannies of the concrete sarcophagus that surrounds Reactor Number 4.

Many of Gaschak's photos could easily pass for snapshots of a pristine wilderness. How could a place so contaminated by nuclear fallout become a seeming paradise for wildlife?

(22–3)

Johnson's interest lies in the flourishing of life in the radioactive zone, terming the entire region a 'radioactive nature preserve' (25). After admitting that the absence of the human element enabled the animals to return and flourish, Johnson writes:

the Exclusion Zone remains the most radioactive environment on Earth. It was so contaminated by nuclear fallout that it won't be completely safe for people to live there for hundreds of years. Hot spots such as the Red Forest will likely be uninhabitable by humans for an even longer period of time. Yet, the Zone's animals don't seem to be aware of the danger all around them.

(25–6)

She then sets out to explore the effects of consuming radioactive plants and life forms on these animals. Initial tests demonstrated no visible effects, leading to the conclusion that the radiation did not harm the animals. Later tests on barn swallows showed they were all radioactive, many showed demonstrable effects like albinism (38). Shortened life spans were also recorded (40). In some species, they found mutations (45).

In a later chapter, Johnson moves out of Chernobyl, into Fukushima. Scientists who had observed Chernobyl went to Fukushima and found the same results:

Timothy Mousseau and Anders Møller have visited Fukushima's exclusion zone every summer to carry out many of the same types of studies they have been doing in Chernobyl. They have seen similar signs of radiation damage in animals from both sites. In places with high levels of radiation, for example, the researchers found significantly fewer than normal numbers of birds, spiders, and insects. And they have spotted barn swallows with patches of white feathers.

(51)

Johnson concludes:

> In short, life has carried on. While scientists continue piecing together the full story of what is happening in the shadow of Reactor Number 4, the animals that creep and scamper and swim and soar through the abandoned lands around it will go on living as best they can in the radioactive kingdom they call home.
>
> (55)

Johnson's text is not in the same category as avowedly anti-nuclear campaign and propaganda texts but can be clubbed with them for what it sets out to do: use images and text to alert us to the non-human consequences of nuclear power, a power that is invisible and yet all-pervasive. The animals 'flourishing' in Chernobyl are life forms that have imbibed and continue to imbibe radiation. Rewilding and refaunation (Grebowicz 59–60) is *radiant* rewilding.

Cornelia Hesse-Honegger, an illustrator for the University of Zurich, collected specimens of insects in areas with nuclear power plants and in regions in direct line of the toxic cloud from Chernobyl: Sweden and Switzerland, and from the Chernobyl region. Over time, Hesse-Honegger collected and drew over 16,000 true bugs from 25 nuclear sites around the world. Later, she took live samples from areas impacted by the Chernobyl cloud, breeding *Drosophila* in her home to observe abnormalities in the offspring. She published a 40-page study 'Malformation of True Bug (Heteroptera): a Phenotype Field Study on the Possible Influence of Artificial Low-Level Radioactivity', with images of these deformed insects as a result of radiation, in the journal *Chemistry & Biodiversity* in 2008. Her other work appeared on Atomic Photographers Guild website (https://atomicphotographers.com/corne lia-hesse-honegger/). The artistic renderings attracted both scientific and public annoyance and fascination, as a Smithsonian essay noted (Thompson).

Hesse-Honegger focuses on the loss of bodily similitude to the normative form of the insect. The deformity implies the presence of something not-quite-life: the toxin. In the process, she captured the transformation of Nature: insects breeding after ingestion of toxins, their genetic material modified and, more seriously, passed on too. As Hesse-Honegger states early in the essay:

> Sensing that Nature was more and more endangered, I gradually developed the notion that mutated laboratory flies were physically rendered prototypes of our destructive behavior, materializing the future of Nature.
>
> (500)

She concludes:

> my field studies show that a significant percentage of European true bugs, living in their specific habitats, are highly disturbed, not only in terms of the actual number of individuals affected, but also regarding the quality and severeness of malformation . . . it confronts us with ethical questions regarding Nature and Life in general. From the scientific point of view, it is necessary 1) to investigate the long-term effects of low-level artificial radiation; 2) to look at the radionuclide-specific effects on plants and animals; and 3) to reconsider the current threshold values for radioactive immission.
>
> (538)

Hesse-Honegger is speaking of a disruption in the order of Nature, as evidenced in the mutated bugs. Her insistence on the ethical questions implies that nuclear energy has interfered with the order(s) in Nature – since the orders in Nature are themselves polyphonous, as Lorraine Daston has argued (2018) – and hence, as biocultural phenomena, needs to be investigated and perhaps stopped.

In his account of the *Lucky Dragon* incident (the US nuclear test, Bravo, at Bikini, 1 March 1954), Ōishi Matashichi reports how contaminated tuna fish from the incident were called 'weeping fish' because each tested specimen had counts of 2000–6000 on the Geiger counter, indicating massive radiation exposure (29).

Addressing a different form of life, Michael Marder writes in *The Chernobyl Herbarium*:

> Chernobyl's human survivors are the scraps of radiation's afterlife, which severely limits life expectancy as a consequence of external and, in many cases, ongoing internal exposures. Plants grown in contaminated soil are, likewise, a finite after life of radiation. Strontium-90 accumulates in vascular vegetal tissues, whereas cesium-137 is distributed throughout a plant, due to its similarity to potassium.
>
> (38)

The photograms composed by Anaïs Tondeur in this volume are, in Marder's terms, 'an emblem – being thrown into, *emballein* – of afterlife' (38).[12] They serve as testimonies and material witnesses to the invisible contamination of the land.

From the above examples, it is clear that the non-human is the silent witness, specimen and spectacle to/of exposure. What Johnson, Marder and Peter Goin set out to do is to *expose the exposure*, so to speak, of the non-human to invisible violence and in the process, further the campaign against nuclearization. Distinct from the aesthetic of the sublime employed to capture large-scale destruction and trauma, the nuclear aesthetic of such artists deals with a reality that does not appear in the frames of what we cognize as real. Representation and aesthetics, they imply, need to move beyond what is visible into a field where there is *nothing to see*. The imperceptible that befalls, and is embodied in, plants and animals (also, of course concrete or geological structures) is, then, the focus of such anti-nuclear arts. In Marder's words, they are 'visible records of an invisible calamity, tracked across the threshold of sight by the power of art' (14). Marder echoes Tim Morton's point in *Hyperobjects*:

> Nuclear radiation is not visible to humans. The nuclear accidents at Chernobyl and Fukushima bathed beings thousands of miles away in unseen alpha, beta, and gamma particles, as radioactive specks floated in air currents across Europe and the Pacific. Days, weeks, months, or years later, some humans die of radiation sickness. Strange mutagenic flowers grow.
>
> (38)

The eerie glow of the plants in *The Chernobyl Herbarium* embodies what may be called after Stacy Alaimo (2016), the 'performance' of exposure. While Alaimo is speaking of an agential act by humans in terms of exposing their (human) vulnerability so as to draw attention to the vulnerability of other life/non-living forms, in this case, I use it to describe the forms of art that document what would otherwise be unexposed in the very act of being exposed to radiation.

Plant life grows in the crevices of concrete structures that are radioactive (Goin, Johnson). Animals consume plants and other creatures and imbibe radioactive nuclides. The plants are captured *with* their 'inner' glow, so to speak. The rhetoric of such art work focuses not on the visible depredations and denudations of the nuclearized land, but on the lushness of life. In a shift away from the traditional frugal economies of environmentalism, these texts show life thriving – 'life finds a way', says Johnson, quoting Ian Malcolm's famous line in Michael Crichton's *Jurassic Park* – in the midst of radiation. There is visible pleasure, Johnson implies, like Marder and Tondeur, in seeing the uninhibited growth of plants and animals. Life then has performed its liveliness in the midst of inhospitable environments. The very notion of an 'Exclusion Zone' is absurd given that nothing is excluded from the radiation that *leaves* the Zone. The art work draws attention to precisely this condition of boundarilessness in the action of radiation. Radiation is the agential act of the non-living upon the living.

Then, the irradiated animals and plants are themselves modes and means of continuing damage, unwittingly of course. Ron Broglio writes of the radioactive rabbit found at the Hanford Nuclear Reservation:

The bunny seems innocuous enough until one realizes that it, being a rabbit, breeds others – and with them more potential carriers of the radioactive conditions of their habitat. If there is one radioactive rabbit then there are other animals out there too. How many?

(unpaginated)

Broglio points to the cause:

As inhuman time moves onward, liquid waste has soaked into the soil. The membranes designed to separate nature and culture have worn down, and radioactive rabbits are the result.

(Broglio, unpaginated)

Referring to the radioactive boar found in Chernobyl and Fukushima, Broglio writes:

The Chernobyl boar aren't just visitors from the past, it turns out. Thanks to the longevity of radiation, they are also visitors from the future.

(Broglio, unpaginated)

This means, the flourishing of life in the 'exclusion zones' is itself a troubling index of a future not yet here, but palpable. In Margret Grebowicz's words: 'The horror of radiation's imperceptibility is that it can show up much later, that you won't know it's happening to you until it's too late and you find yourself the subject of a monstrous birth' (63). The herbarium and the refaunation are to be read as signs of the monstrous future of the planet where toxins of the disasters and tests linger *through life's very acts of survival* (breeding).

These texts are countable as anti-nuclear texts precisely because they all signal the horror of a nuclearized planet in which there is no possibility of an 'Exclusion Zone'.

Extinction Iconography and the Nuclear Cosmogram

Nevil Shute's *On the Beach* (1957) set in Australia, specifically Melbourne, depicts the last humans on earth as the radiation sickness wipes out city after city, sweeping downward from Europe. Rather than die painfully and slowly, the people have the option of taking suicide pills, supplied by the Australian government. Looking forward to Alan Weisman's bestselling thought-experiment, *The World without Us* (2007), Shute's John Osborne says:

"It's not the end of the world at all," he said. "It's only the end of us. The world will go on just the same, only we shan't be in it. I dare say it will get along all right without us."

(unpaginated)

Shute's novel and Osborne's comment capture the 'near/extinction' – I use this awkwardly structured phrasing to suggest that it is not always total extinction but *near* or *almost* extinction, but also *close to* extinction that the discourses imply – discourse around the initiated in Mary Shelley's *The Last Man* (1826). However, the anxieties around *nuclear* energy and its extinction emerged in full-fledged form, as Daniel Cordle argues (4), first in the 1940s and 1950s, although the nuclear winter theory and related fears emerged after Three-Mile Island and Chernobyl in the late 1970s.

Near/Extinction Subjects

There is extinction discourse of the kind visible in Shelley and Shute but also a *near*-extinction discourse wherein the literature envisages an almost-apocalypse survived by a handful of people who have to re-populate the earth (or, in some science fiction texts, other worlds/planets). In this section, I merge these two interrelated themes to speak of a near/extinction discourse in the novels published in the immediate aftermath of the Hiroshima–Nagasaki bombings and the first wave of nuclear testing: the 1940s and 1950s.[13]

In Shute's novel, the imminence of human death is described thus:

> We've all got to die one day, some sooner and some later. The trouble always has been that you're never ready, because you don't know when it's coming. Well, now we do know, and there's nothing to be done about it. I kind of like that. I kind of like the thought that I'll be fit and well up till the end of August and then – home. I'd rather have it that way than go on as a sick man from when I'm seventy to when I'm ninety.
>
> (unpaginated)

The elegiac-tragic tone of the novel apart – a characteristic of extinction discourse, as Ursula Heise has observed (2016) – Shute's comment cited above captures one of the key features of the first wave of near-extinction nuclear literature: death-as-imminent is coterminous in nuclear anxieties with near-extinction as *immanent* (as Frank Kermode would propose about a certain kind of apocalyptic thought and tragedy in *The Sense of an Ending*). With nuclear technology and nuclear war, near/extinction is built into the very employment of this power, everyday life and the cultural imaginary.

Central to the near/extinction discourse of the 1940s and 1950s is the immanence of mutual or partial destruction. American novels such as Philip Wylie's *Tomorrow* (1954) depict the nuclear subjects of two small towns in the USA, Green Prairie and River City. These subjects believe nuclear bombardment by the Russians is imminent and death from radiation sickness is immanent in their lives in a nuclearized world. The in/security discourse that dovetails into the near/extinction discourse drives the community to create Civil Defence structures and regular drills in the *eventuality* of a nuclear attack. In the novel, Charles is annoyed that not all members of the entire take seriously the imminence and immanence of destruction in the technologies possessed by both the USA and Soviet Union. Charles believes that

> this ingrained sense that River City would always be there because it had always been there, the emotional identification with the imme-diate here and the refusal even to look at the hard and horrible face of tomorrow yonder, annoyed Charles . . .
>
> (Wylie unpaginated)

The future monstrosity of nuclear war and extinction is something that must inform the present, Wylie's Charles seems to indicate.

As Kermode argues: 'we think in terms of crisis rather than temporal ends' (30) and the 'weight of End-feeling' is 'throw[n] on to the moment, the crisis' (25). This is what I term the feeling and belief of immanent near-extinction, and in works like Wylie's, the discourse is employed to con-struct a subject in a constant state of nuclear fear. In Wylie's *Tomorrow!* Coley Borden's editorial opens with a detailed exposition of nuclear fear and the imminence/immanence of nuclear power's assured destruction:

> Ten years ago and more, this nation hurled upon its Jap foe a new weapon, a weapon cunningly contrived from the secrets of the sun. Since that day the world has lived in terror.
>
> (unpaginated)

Here, in the opening lines of his editorial, Borden does not identify American nuclear fears and subjectivity, but expands it to include the world. He builds on the theme of a universal culture of fear:

> Every year, every month, every hour, terror has grown. It is terror compounded of every fear. Fear of War. Fear of Defeat. Fear of Slavery. These fears are great, but they are common to humanity.
>
> (unpaginated)

About America he says:

> They are then puppets of their terror. And it is as such puppets that we Americans have acted for ten years, and more.
>
> (unpaginated)

Then:

> We, the people of the United States of America, have refused for more than a decade to face our real fear. We know our world could end.
>
> (unpaginated)

He argues that Americans have prepared themselves for cyclones and fires, but *'what of the peril of world's end?'* he asks (emphasis in original). He concludes with:

> Now, the sands of a decade and more have run out. We cannot challenge without venturing the world's end. Quite possibly our death notice is written, a few months or years farther along on the track of this wretched planet. Then, perhaps, our flight from freedom will get the globe rent into hot flinders, atomized gas.
>
> (unpaginated)

Borden's near-rant about American unpreparedness in the face of imminent/immanent nuclear destruction – 'the rage of radioactivity, the blast of neutrons' (unpaginated) – is, at the very beginning of the Cold War period, a summary of the discourse of near/extinction, which is employed to justify heightened weaponization and securitization, even as it thrives on the in/security discourse of the age. The rant is also about the world at large.

The imminence/immanence state of being does have a temporal dimension in the near/extinction discourse – and this has to do with 'how these imaginaries of the future in turn shape constructions of the present and the past' (Kaplan 12). In Borden's extended meditation on the imminence/immanence of assured destruction, he employs the imagination of *future* destruction to reflect on the way America *has been* over the years. That is, the imminent destruction from Soviet nuclear attacks are, in Borden's analysis, immanent in the way America has behaved in the past, beginning with the *American* bombing of Japan. Borden is sketching what Ann Kaplan argues, are 'pretraumatic scenarios' of 'futurist disaster narratives' (3) but constantly embedding them in the past and the present. In the process, he calls for greater attention to the present day measures that need to be taken. The nuclear subject who emerges in this pretraumatic scenario, in other words, needs to re-form her/his self today: for 'imagined future selves have an impact on one's current view of self' (Kaplan 7).

Near/Extinction Iconography

In John Wyndham's post-apocalyptic novel *Re-birth* (1955), the remainder of 'civilization' – set somewhere in Canada – is back to an old-world existence, eking out a living through agriculture and in village communities (we later discover that an advanced civilization lives in 'Zealand'): 'Ours was no longer a frontier region. Hard work and sacrifice had produced a stability of stock and crops' (unpaginated).[14]

In this community where humanity struggles to rebuild itself, children are born with deformities but also strange powers, such as telepathy. These are called 'deviations' – mutations in human beings exposed to radiation. The children are brought up to abhor 'deviations' and warned constantly: 'Watch Thou for the Mutant!' The mutants are nearly human:

> although they were really Deviations they often looked quite like ordinary human people, if nothing had gone too much wrong with them.
>
> (unpaginated)

David, the boy-narrator of the tale, expects some strange sights when he encounters the people from the 'Fringes':

> It was a bit disappointing at first sight. The tales about the Fringes had led me to expect creatures with two heads, or fur all over, or half a dozen arms and legs. Instead, they seemed at first glance to be just two ordinary men with beards – though unusually dirty, and with very ragged clothes.
>
> (unpaginated)

There is then a pretraumatic envisioning of the end of humanity as we know it and the rise of a new species in Wyndham.

The mutant in the Wyndham text is an extinction icon. Extinction icons are constructed through social discourses and imaginaries, argue Samuel Turvey and Anthony Cheke (2008). In cases such as the Dodo's, they note, 'it is frequently also popularly assumed that those who were responsible for its extinction were aware of what they had done' (149). As Wyndham presents it, the 'deviations' are the long-term consequences of the nuclearized world of the future. They capture at once the end of humanity as we know it, and the rise of a new species of humans.[15] They are, then, both extinction and redemption icons. As the Zealand woman (identified as just that) comments at the end of the novel:

> There have been lords of life before, you know. Did you ever hear of the great lizards? When the time came for them to be superseded they had to pass away.
>
> Sometime there will come a day when we ourselves shall have to give place to a new thing. Very certainly we shall struggle against the inevitable just as these remnants of the Old People do. We shall try with all our strength to grind it back into the earth from which it is emerging, for treachery to one's own species must always seem a crime. We shall force it to prove itself, and when it does, we shall go; as, by the same process, these are going . . . ours is a superior variant, and we are only just beginning.
>
> (unpaginated)

The 'beginning' implies the rise of a new human species, in a world altered irrevocably by nuclear energy. This is at once extinction and renewal.

The reprisals against the mutants is the struggle of the 'norms' (as the 'normal' humans are called in the novel) against the extinction-redemption of the human race itself. When Wyndham's character draws an analogy between the extinction of the lizards (the dinosaurs, one assumes) and that of the current 'model' of humanity, she is gesturing at the human responsibility for extinction and inadvertent redemption: it is man-made technology that produced wars and radiation which in turn induced the mutations and thus a new race/model of the human.

Wyndham's vision has less to do with extinction *per se* than with the cycle of life, survival and extinction – the saecular narrative that marks the apocalyptic vision (Kermode) where the civilization is renewed from time to time. Saeculae

> bear the weight of our anxieties and hopes; they are "intemporal/ but we project them onto history, making it "a perpetual calendar of human anxiety." They help us to find ends and beginnings. They explain our senescence, our renovations . . .
>
> (Kermode 11)

Kermode identifies in the apocalyptic narrative, a certain myth of transition, where there is a 'period which does not properly belong either to the End or to the *saeculum* preceding it' (12). The mood captured in Wyndham, as in the other nuclear narratives, is of a pattern, a cycle and a transition:

> The mood of *fin de siecle* is confronted by a harsh historical *finis saeculi*. There is something satisfying about it, some confirmation of the Tightness of the patterns we impose . . . the anxiety reflected by the *fin de siecle* is perpetual, and people don't wait for centuries to end before they express it. Any date can be justified on some calculation or other.
>
> (Kermode 98)

The saecular narrative around nuclear power may be said to have begun with the Trinity test itself. Enrico Fermi, for example, wondered before the test 'whether or not the bomb would ignite the atmosphere, and if so, whether it would merely destroy New Mexico or destroy the world' (cited in Groves 296). The bomb's testing gives many the scope of looking at mankind's future. George Kistiakowsky exclaimed after watching the Trinity Test:

> This is the nearest thing to doomsday that one could possibly imagine. I am sure that at the end of the world – in the last millisecond of the Earth's existence – the last human will see what we saw.
>
> (Kistiakowsky unpaginated)

William Laurence, the only journalist present at the Trinity test, writes: 'the Atomic Age began at exactly 5.30 Mountain War Time on the morning of July 16, 1945 . . . ' (22). Later, he would call it by other names: 'a new cultural age, the Age of Atomics, or the Age of Nucleonics' (51). Richard Rhodes in his *The Making of the Atomic Bomb* cites Emilio Segrè, who expects the bomb to 'set fire to the atmosphere and thus finish the earth' (673). He also cites Oppenheimer speaking later about Trinity: 'We knew that it was a new world' (676).

Doomsday and the birth of a new age merge in the rhetoric of Trinity, and Wyndham's fiction was repeating the saecular narrative of the nuclear tests, of the rise and fall of a human civilization, and a fall, engineered by man's own hands. It could then be argued that the Trinity test and the first mushroom cloud over New Mexico also served for the scientists and others who witnessed it, as an extinction icon: *the bomb heralded both, a new age, and the potential to annihilate ourselves as a race.* Humanity had just been reinvented as nuclear subjects by its own doings.

Extinction iconography, as seen in these texts, may very well have its own symbolism, a nuclear cosmogram, no less.

The Nuclear Cosmogram

The symbol of radiation hazard is now a common visual vocabulary for the planet. Used across the nuclear sites in the world, the symbol is indeed a symbol of planetary precarity. It is a cosmogram.

A cosmogram is 'a text that results in a concrete practice and set of objects, which weave together a complete inventory or map of the world' (John Tresch, cited in Gabrielle Hecht 102). In Gabrielle Hecht's gloss, 'Cosmograms concretize: they offer a set of practices, of rituals to enact participation in the world' (107). The symbol for radiation is, I suggest, a cosmogram that enables an understanding of the world, our participation

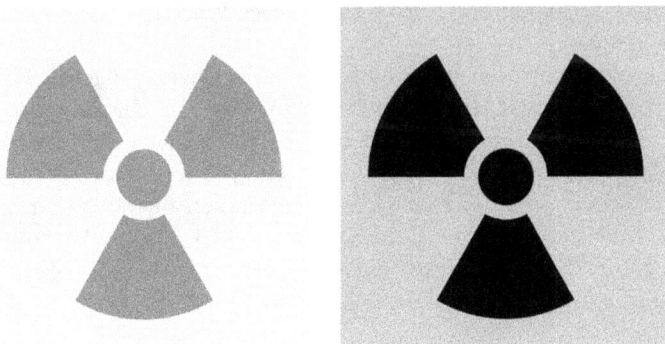

Figure 5.3 Radiation hazard symbol.

in it, and a set of practices – all of which are to do with the presence of a nuclear threat.

The ritual posting of the symbol at multiple spots around the world brings the world together into the orbit of nuclearization and its attendant risks, potential and threat. It cautions people against specific areas.[16] The three propeller blades were meant to illustrate the radiation activity of the atom at the centre.

But it could also signify the earth in the middle hemmed in by activity of some kind, from all three sides. That is, it need not be the centrifugal (radiating outward from the centre) but a centripetal movement of the danger from the outside enveloping or closing in.

As a cosmogram, the radiation hazard symbol is literally then a map of the planet, perhaps photographed or drawn from an aerial view and therefore the nuclear equivalent of the famous Blue Marble photograph of NASA's Apollo 17. The view from above in photography, write Denis Cosgrove and William Fox,

> allows us to plan ahead, to place ourselves in the larger context of the world and map out a course in both space and time. The fundamental utility of the aerial view is why we are so attracted to the examples in the museum – we can't help looking at a view that we know carries such important information.
>
> (11–12)

They elaborate:

> aerial photography does remarkably well is to reveal and even create pattern at varying scales on the earth's surface. Unless viewed stereoscopically or taken from an oblique angle in sharp shadow, it also flattens the image, emphasizing surface.
>
> (100)

The earth in the middle appears shrunk ('not to scale'!) and overwhelmed by the blades surrounding it from three sides. It informs us that the earth is *surrounded*.

The second cosmogram would be the Doomsday Clock. Designed by Martyl Langsdorf, it was unveiled in 1947 by the *Bulletin of the Atomic Scientists* (a group founded by Einstein and others involved in the Manhattan Project), whose first issue appeared four months after Hiroshima–Nagasaki. The Clock has been taken to indicate how close we are to disaster/midnight. Others claim: 'the Clock is meant to indicate the current level of risk facing humanity' (Beard). SJ Beard, who works in the field of existential risk analysis, writes:

> Its [the Clock's] goal is not to tell us how big the risk facing humanity is, but how well we are doing at responding to that risk. For instance,

in 1962 the Cuban Missile crisis is generally agreed to have been the closest the world ever came to nuclear war, but its occurrence did not move the Clock at all. On the other hand, the 1963 Partial Nuclear Test Ban Treaty saw the Clock's hands shifted back from midnight an entire five minutes.

(2022)

(The Clock, originally designed to reflect the threat from nuclearization, has included climate change in its calculations since 2007.)

The Clock is a symbol and is not a scientific or quantitative assessment of the risk the planet is embedded in. Rachel Bronson, the *Bulletin* President and the CEO explains:

When the clock is at midnight, that means there's been some sort of nuclear exchange or catastrophic climate change that's wiped out humanity," she said. "So we never really want to get there and we won't know it when we do."

(Marples and Ramirez 2022)

The Clock signals how close we are to annihilation and the efforts we, as a species, are making to delay the catastrophic event. Setting the clock back or forward, then, constitutes an awareness and an action plan vis-à-vis planetary annihilation.

If the regular clock *shows* the passage of time in the form of the moving hands, the Doomsday Clock shows the time *left* for us as a planet. Then, the Doomsday Clock as a cosmogram is not, unlike a traditional watch/clock, pointing to the local time, or the time at a specific place. Rather, it symbolizes something as unquantifiable and even unimaginable as planetary time: the time of the earth itself and all the matter on it.

Taken together, the Doomsday Clock and the radiation symbol signal a planetarity: that wherever you are, your future (time) remains linked to the rest of the planet.

It is a planet in peril.

Representing the threat to the planet from nuclear power takes multiple modes, and this chapter has examined a select number. It has identified discourses of anti-nuclear sentiment and caution, in the form of literary and artistic texts, but also campaign rhetoric and early advocacy texts from scientists and statesmen.

The extinction iconography is integral to the discourse of planetary precarity and has been found in literary texts right from the 1950s, as this chapter has shown. What emerges from even the most modest scrutiny is a sense of place, but also that it is indeed, in the face of nuclear power's omniscience – in space and time, since it extends into the future – 'one world or none'.

Notes

1 There is another form of precarity in literature dealing with nuclear tests and their aftermath. In Spitz's *Island of Shattered Dreams*, the island is transformed after the arrival of the nuclear test base. Spitz writes of the loss of the culture of the place as a precarity:

> In twenty years Ruahine has gone through the kinds of changes that took place over two millennia in Metropolitan France. The white man's craziness has once again struck this quiet island, maddening its inhabitants who haven't been able to protect themselves against the devastating torrent of western modernity. How could they have protected themselves, swept away by time suddenly rushing on, drowned in values so different from their own, in a world where profit and corruption rule people's minds?
>
> (129)

The world of the island is endangered not just from the nuclear base but culturally: the arrival of consumerism, European modes of life and the erosion of native habits and customs. 'The Ma'ohi people are victims of an insatiable appetite for imported products and consume everything without thinking', writes Spitz, thus capturing the collapse of local cultures (130).

2 Although this is not the subject of the current chapter, it is interesting to note how anti-nuclear movements also reconfigured the nuclear scientist. In the words of Spencer Weart:

> For antinuclear people, the ambiguous structure that I have described earlier in this book – scientist and victim, monster and hero, promiscuously interchanging places – resolved into a simple pattern. They saw all atomic affairs through the lens of a simple bipolar structure that might be written in shorthand as *authority/subject*. On one side stands the scientist or technologist with his dangerous devices, the domineering male (especially a government, industry, or military official), the authoritarian father, the entire generalized threat from an overregulated technological society. On the other side stands the guinea pig, the enslaved worker, the dominated woman, the rejected child, the individual crushed by modern society.
>
> (Weart 211–2)

3 The American public discourse around nuclear waste has not remained uniform or homogenous through the twentieth and twenty-first centuries. In the first decades (1950–1960s), the emphasis was on the beneficial effects of nuclear power. But after 1970, the paradigm shifted towards the harm potential of nuclear power and nuclear waste. For a study of this history of American perceptions and public discourse in the 1950–2009 period, see Pajo (2016). For studies of selective region-wise public responses to siting nuclear waste disposal, see Dunlap et al. (1993).

4 The film also traces a history of incremental destruction due to newer and newer weapons in war: the spear killed one man, the cannons killed 88 and so on, in order to argue that the scale of destruction has always been rising.

5 Gabrielle Hecht in her essay, 'The Power of Nuclear Things' (2010) traces the genealogy of uranium being transformed from mineral to a 'nuclear thing' or 'nuclear material'. Hecht argues that the point at which uranium becomes weapons-usable is dependent on 'time, place, purpose, and market' (2). This

process is also linked, notes Hecht, to the way in which Africa – the source of much uranium for the USA, Europe and Japan – is itself imagined and constructed in the nuclear discourse.

6 For a study of the nuclear stockpile and stewardship rhetoric, see Taylor and Hendry (2008).

7 'Preoccupied', the *OED* tells us, is also to 'engross', 'dominate', 'distract'.

8 Even the *hibakushas*, argue some commentators and novelists, were erased from memories and the land. In James George's *Ocean Roads*, Akiko tells Caleb:

> At one end of the building, there was a ward just for them. Even those who had survived had had their names wiped away, by my people, by the Americans. The great, collective forgetting. United only by their bandages.
>
> (328)

Robert Jay Lifton and Greg Mitchell would state in the Introduction to *Hiroshima in America: A Half Century of Denial*:

> Hidden from the beginning, Hiroshima quickly disappeared into the depths of American awareness…Whatever our avoidance and numbing, Americans remained haunted by the atomic bombings, all the more so because they remain obscure, distant, and mysterious, and because they were our doing.
>
> (xv)

9 That the concept of *terra nullius* adversely impacts indigenous people is cited by the UN itself. See 'Doctrines of Dispossession' (www.un.org/WCAR/e-kit/backgrounder4.htm)

10 In 1989, the Department of Energy abandoned its effort to assess the technical suitability of the Yucca Mountain as a potential site for the nation's only civilian high-level nuclear waste repository.

11 Similar instances of life forms thriving in regions of nuclear testing, such as Bikini Atoll, have often appeared in the mass media. See, for example, Keating (2017) and Sperling (2020). Ōishi Matashichi notes how, even in Japan, in the early 1950s, 'many articles on the peaceful uses of atomic energy' were published in the Japanese press, despite memories of Hiroshima–Nagasaki being raw in Japanese minds (68). As Lauren Redniss puts it in her graphic biography of Marie Curie, the return of animal life to the Exclusion Zone in Chernobyl has caused people to 'announce the creation of an inadvertent wildlife refuge: an accidental Eden' (113). This myth has been debunked by Tim Mousseau's celebrated work on the animal life of Chernobyl and Fukushima and the impact of radiation on them.

12 The photograms were 'created through the direct imprints of radioactive herbarium specimens, grown in the soil of "the exclusion zone" by Martin Hajduch of the Institute of Plant Genetics and Biotechnology at the Slovak Academy of Sciences and arranged on photosensitive paper' (Marder and Tondeur 11).

13 The apocalyptic novel has mushroomed – pun intended – in the 1980s and 1990s, in both popular and high literary modes, from Stephen King's *The Stand* through P.D James' *The Children of Men* to Octavia Butler's *Xenogenesis* trilogy and Margaret Atwood's *MaddAddam* trilogy.

14 'Re-birth' is the alternative title for *The Chrysalids*.

15 Here, it is pertinent to mention that Octavia Butler in *Xenogenesis* also imagines a future with a new species of humans – except that in this case, she sees this emergence as a result of the symbiosis of human survivors on earth with an alien life form, the Oankali.

16 On the radiation hazard sign's origins, see Lloyd Stephens and Rosemary Barrett, 'A Brief History of a 20th Century Danger Sign' (1978).

Bibliography and Filmography

Agyeya [SH Vatsyayan]. 'Hiroshima'. Trans. Agyeya. In Kailash Vajpayi (ed) *An Anthology of Modern Indian Poetry*. New Delhi: Rupa, 2000. www.cse.iitk. ac.in/users/amit/books/vajpeyi-2000-anthology-of-modern.html. 3 May 2022.

A Hard Rain. Director David Bradbury. New South Wales, Australia: Frontline Films, 2007.

Alaimo, Stacy. *Exposed: Environmental Politics and Pleasures in Posthuman Times*. Minneapolis and London: University of Minnesota Press, 2016.

Alexievich, Svetlana. *Chernobyl Prayer: A Chronicle of the Future*. Trans. Anna Gunin and Arch Tait. New Delhi: Penguin, 2016.

Allen, Margaret. 'Homely Stories and the Ideological Work of "Terra Nullius"', *Journal of Australian Studies* 27.79 (2003): 105–15.

Alpers, Svetlana. 'The Studio, the Laboratory, and the Vexations of Art', in Caroline A. Jones and Peter Galison (eds) *Picturing Science, Producing Art*. London: Routledge, 1998. 401–17.

Andren, Mats. 'Atomic War or World Peace Order? Karl Jaspers, Denis de Rougemont, Bertrand Russell', *Global Intellectual History* 7.4 (2022): 784–800.

Andrews, Hannah. 'Recitation, Quotation, Interpretation: Adapting the Ouevre in Poet Biopics', *Adaptation* 6.3 (2013): 365–83.

Anolik, Ruth Bienstock. 'Introduction: Diagnosing Demons: Creating and Disabling the Discourse of Difference in the Gothic Text', in Ruth Bienstock Anolik (ed) *Demons of the Body and Mind: Essays on Disability in Gothic Literature*. Jefferson, North Carolina and London: McFarland, 2010. 1–19.

Arribert-Narce, Fabien. 'Narrating Fukushima: The Genre of "Notes" as a Literary Response to the 3/11 Triple Disaster in Hideo Furukawa's *Horses, Horses, in the End the Light Remains Pure* (2011) and Michaël Ferrier's *Fukushima: Récit d'un désastre* (2012)'. *a/b: Auto/Biography Studies*. 2021. doi:10.1080/08989575.2021.1940634.

Aubin, David and Charlotte Bigg. 'Neither Genius nor Context Incarnate: Norman Lockyer, Jules Janssen and the Astrophysical Self', in Thomas Söderqvist (ed) *The History and Poetics of Scientific Biography*. Aldershot: Ashgate, 2007. 51–70.

Baldick, Chris. 'Introduction', in Baldick (ed) *The Oxford Book of Gothic Tales*. New York: Oxford University Press, 1992. xi–xxiii.

Banco, Lindsey Michael. 'The Biographies of J. Robert Oppenheimer: Desert Saint or Destroyer of Worlds', *Biography* 35. 3 (2012): 492–515.

Bastardoz, Nicholas. 'Signaling Charisma', in José Pedro Zúquete (ed) *Routledge International Handbook of Charisma*. Oxford: Routledge, 2021. 313–23.

Beard, S.J. 'How to Read the Doomsday Clock'. *BBC*. 20 January 2022. www.
bbc.com/future/article/20220119-how-to-read-the-doomsday-clock.   30
March 2022.

Bernstein, Barton J. 'In the Matter of J. Robert Oppenheimer', *Historical Studies
in the Physical Sciences* 12.2 (1982): 195–252.

Bernstein, Jeremy. *Oppenheimer: Portrait of an Enigma*. Chicago: Ivan
R. Dee, 2004.

Bethe, Hans. 'J. Robert Oppenheimer'. In *National Academy of Sciences:
Biographical Memoirs, Vol. 71*. Washington, DC: National Academy Press,
1997. 175–220. https://nap.nationalacademies.org/read/5737/chapter/12. 3
April 2022.

Bird, Kai and Martin J Sherwin. *American Prometheus: The Triumph and Tragedy
of J. Robert Oppenheimer*. New York: Vintage, 2005.

Biswas, Shampa. *Nuclear Desire: Power and the Postcolonial Nuclear Order* .
Minneapolis and London: University of Minnesota Press, 2014.

Blernoff, Suzannah. 'The Corporeal Sublime', *Australian and New Zealand
Journal of Art* 3.1 (2002): 60–75.

Bohr, Neils. 'Open Letter to the United Nations'. 1950. Archived at www.ato
micheritage.org/key-documents/bohr-letter-un. 12 March 2022.

Borch, Merete. 'Rethinking the Origins of *terra nullius*', *Australian Historical
Studies* 32.117 (2001): 222–39.

Borovoi, Alexander A. *My Chernobyl: The Human Story of a Scientist and the
Nuclear Plant Catastrophe*. Trans. Julya Borovoi. Portsmouth: Piscataqua
Press, 2017.

Bosch, Mineke. 'Persona and the Performance of Identity: Parallel Developments
in the Biographical Historiography of Science and Gender, and the
Related Uses of Self Narrative'. *L' homme: Zeitschrift für feministische
Geschichtswissenschaft,* 24.2 (2013): 11–22. Gender Open Repositorium.
www.genderopen.de/bitstream/handle/25595/1054/lhomme.2013.24.2.11.
pdf%20Bosch.%20Mineke.pdf?sequence=1&isAllowed=y. 26 Dec. 2021.

———. 'Scholarly Personae and Twentieth-Century Historians: Explorations of
a Concept', *BMGN-Low Countries Historical Review* 131. 4 (2016): 33–54.

Boyer, Paul. *Fallout: A Historian Reflects on America's Half-Century Encounter
with Nuclear Weapons*. Columbus: Ohio State University Press, 1998.

Bradley, John. Ed. *Atomic Ghost: Poets Respond to the Nuclear Age*. Minneapolis:
Coffee House, 1995.

Broderick, Mick. Ed. *Hibakusha Cinema: Hiroshima, Nagasaki and the Nuclear
Image in Japanese Film*. London and New York: Routledge, 1996.

Brodie, Janet Farrell. 'Radiation Secrecy and Censorship after Hiroshima and
Nagasaki', *Journal of Social History* 48.4 (2015): 842–64.

Broglio, Ron. 'The Creatures That Remember Chernobyl'. *The Atlantic*. 26
April 2016. www.theatlantic.com/science/archive/2016/04/the-creatures-that-
remember-chernobyl/479652/. 15 March 2022.

Broinowski, Adam. 'The Atomic Gaze and *Ankoku Butoh* in Post-war Japan', in
N.A.J. Taylor and Robert Jacobs (eds) *Reimagining Hiroshima and Nagasaki:
Nuclear Humanities in the Post-Cold War*. Oxford: Routledge, 2018. 91–107.

Brown, Kate. *Manual For Survival: A Chernobyl Guide to the Future*. New York:
W.W. Norton, 2019. eBook.

Bruzzaniti, Giuseppe. *Enrico Fermi: The Obedient Genius*. Trans. Ugo Bruzzo.
New York: Springer, 2016.

Buchan, Bruce. 'Traffick of Empire: Trade, Treaty, and *Terra nullius* in Australia and North America, 1750-1800', *History Compass* 5.2 (2007): 386–405.

Buck, Alice L. *A History of the Atomic Energy Commission*. Washington: US Department of Atomic Energy, 1983.

Buddha Weeps in Jadugoda. Director Shri Prakash, 1999.

Burchett, Wilfred. 'The Atomic Plague', in George Burchett and Nick Shimmin (eds) *Rebel Journalism: The Writings of Wilfred Burchett*. Cambridge: Cambridge University Press, 2007. https://hibakushastories.org/wp-content/uploads/2018/09/The-Atomic-Plague.pdf. 3 March 2022.

Burgess, Victoria. 'Down Winder from the Mushroom Clouds'. 2012. https://collections.lib.utah.edu/details?id=1248578&facet_setname_s=uum_dua. 3 April 2022.

Byerly, Alison. ' "A Prodigious Map Beneath His Feet": Virtual Travel and The Panoramic Perspective', *Nineteenth-Century Contexts* 29.2–3 (2008): 151–68.

Cantor, Geoffrey. 'The Scientist as Hero: Public Images of Michael Faraday', in Michael Shortland and Richard Yeo (eds) *Telling Lives in Science: Essays on Scientific Biography*. Cambridge: Cambridge University Press, 1996. 171–94.

Chakrabarty, Dipesh. 'The Politics of Climate Change Is More Than the Politics of Capitalism', *Theory, Culture & Society* 34.2–3 (2017): 25–37.

Chaplin, Joyce E. and Darrin M. McMahon. 'Introduction', in Joyce E. Chaplin and Darrin M. McMahoon (eds) *Genealogies of Genius*. London: Palgrave Macmillan, 2016. 1–10.

Chute, Hillary L. *Disaster Drawn: Visual Witness, Comics and Documentary Form*. Cambridge, MA: Belknap-Harvard University Press, 2016.

Cixous, Hélène. 1976. 'Fiction and Its Fantoms: A Reading of Freud's Das Unheimliche (The "Uncanny")', *New Literary History* 7: 525–48.

Clark, John. 'We Were Trapped by Radioactive Fallout'. *The Wetokian*. Fall 1999. www.sonicbomb.com/content/atomic/docs/trapped_by_radioactive_fallout.pdf. 19 Jan. 2022.

Cochrane, Tom. 'The Emotional Experience of the Sublime', *Canadian Journal of Philosophy* 42.2 (2012): 125–48.

Colebrooke, Claire. *The Death of the PostHuman: Essays on Extinction, Volume One*. London: Open Humanities Press, 2014.

Cordle, Daniel. *Late Cold War Literature and Culture: The Nuclear 1980s*. London: Palgrave Macmillan, 2017.

Cosgrove, Denis and William L. Fox. *Photography and Flight*. London: Reaktion, 2010.

Couser, G. Thomas. *Signifying Bodies: Disability in Contemporary Lifewriting*. University of Michigan Press, 2012.

Cunningham, David. 'How the Sublime Became "Now": Time, Modernity, and Aesthetics in Lyotard's Rewriting of Kant', *Contemporary Issues in Aesthetics* 8.3 (2004): 549–71.

Curie, Eve. *Madame Curie*. Trans. Vincent Sheean. New York: Doubleday, 1939.

Daston, Lorraine and H. Otto Sibum. 'Introduction: Scientific Personae and Their Histories', *Science in Context* 16.1–2 (2003): 1–8.

Daston, Lorraine. *Against Nature*. Cambridge, MA: MIT Press, 2018.

De Boer, Bas, Hedwig te Molder and Peter-Paul Verbeek. 'Understanding Science-in-the-Making by Letting Scientific Instruments Speak: From Semiotics to Postphenomenology', *Social Studies of Science* 51.3 (2021): 392–413.

DeFalco, Amelia. *Uncanny Subjects: Aging in Contemporary Narrative.* Columbus, OH: Ohio State University Press, 2010.

DeLoughrey, Elizabeth. 'Heliotropes: Solar Ecologies and Pacific Radiations', in Elizabeth DeLoughrey and George B. Handley (eds) *Postcolonial Ecologies.* Oxford: Oxford University Press, 2011. 235–53.

———. 'The Myth of Isolates: Ecosystem Ecologies in the Nuclear Pacific', *Cultural Geographies* 20.1 (2012): 167–84.

Derrida, Jacques. 'No Apocalypse, Not Now: (Full Speed Ahead, Seven Missiles, Seven Missives)'. Trans. Catherine Porter and Philip Lewis. *Diacritics* 14.2 (1984): 20–31.

———. 'Demeure: Fiction and Testimony', in Maurice Blanchot (ed) *The Instant of My Death/Demeure: Fiction and Testimony.* Trans. Elizabeth Rottenberg. Stanford: Stanford University Press, 2000.13–103.

Dibblin, Jane. *Day of Two Suns: US Nuclear Testing and the Pacific Islanders.* New York: New Amsterdam, 1990.

DiNitto, Rachel. *Fukushima Fiction: The Literary Landscape of Japan's Triple Disaster.* Honolulu: University of Hawaii Press, 2019.

Dirt Cheap 30 Years On: The Story of Uranium Mining in Kakadu. Gundjeihmi Aboriginal Corporation in association with The Environment Centre NT, 2011.

Dolph Briscoe Center. *Flash of Light, Wall of Fire: Japanese Photographs Documenting the Atomic Bombings of Hiroshima and Nagasaki.* Austin: University of Texas Press, 2020.

Dower, John. 'War, Peace, and Beauty: The Art of Iri Maruki and Toshi Maruki', in John Dower and John Junkerman (eds) *The Hiroshima Murals: The Art of Iri Maruki and Toshi Maruki.* Tokyo: Kodansha International, 1985. 9–26.

Dower, John and John Junkerman. Ed. *The Hiroshima Murals: The Art of Iri Maruki and Toshi Maruki.* Tokyo: Kodansha International, 1985.

Dunlap, Riley E., Michael E. Kraft, and Eugene A. Rosa. Eds. *Public Reactions to Nuclear Waste: Citizens' Views of Repository Siting.* Durham: Duke University Press, 1993.

Duro, Paul. 'Observations on the Burkean Sublime', *Word & Image* 29.1 (2013): 40–58.

Edensor, Tim. *Industrial Ruins: Spaces, Aesthetics, and Materiality.* Oxford: Berg, 2005.

Edwards, Paul N. 'Entangled Histories: Climate Science and Nuclear Weapons Research', *Bulletin of the Atomic Scientists* 28.4 (2012): 28–40.

Effects of the Atomic Bombs: Report of the British Mission to Japan. London: For the Home Office and the Air Ministry by His Majesty's Stationery Office, 1946.

Einstein, Albert. 'The Way Out', in Dexter Masters and Katharine Way (eds) *One World or None.* New York: McGraw-Hill, 1946. 76–7.

Endres, Danielle. 'The Rhetoric of Nuclear Colonialism: Rhetorical Exclusion of American Indian Arguments in the Yucca Mountain Nuclear Waste Siting Decision', *Communication and Critical/Cultural Studies* 6.1 (2009): 39–60.

Endres, Danielle and Samantha Senda-Cook. 'Location Matters: The Rhetoric of Place in Protest', *Quarterly Journal of Speech* 97.3 (2011): 257–82.

Engler, Mira. 'Repulsive Matter: Landscapes of Waste in the American Middle-Class Residential Domain', *Landscape Journal* 16.1 (1997): 60–79.

Eubanks, Charlotte. 'The Mirror of Memory: Constructions of Hell in the Marukis' Nuclear Murals', *PMLA* 124. 5 (2009): 1614–31.

Fahy, Declan and Bruce Lewenstein. 'Scientists in Popular Culture: The Making of Celebrities', in Massimiano Bucchi and Brian Trench (eds) *Routledge Handbook of Public Communication of Science and Technology*. London: Routledge, 2021. 3rd ed. www.routledgehandbooks.com/doi/10.4324/978100 3039242-3-3. 25 December 2021.

Fara, Patricia. 'Framing the Evidence: Scientific Biography and Portraiture', in Thomas Söderqvist (ed) *The History and Poetics of Scientific Biography*. Aldershot: Ashgate, 2007. 71–93.

Federation of Atomic Scientists. 'Survival Is at Stake', in Dexter Masters and Katharine Way (eds) *One World Or None*. New York: McGraw-Hill, 1946. 78–9.

Ferguson, Frances. 'The Nuclear Sublime', *Diacritics* 14.2 (1984): 4–10.

Fetter-Vorm, Jonathan. *Trinity: A Graphic History of the First Atomic Bomb*. New York: Hill and Wang, 2012.

Fiege, Mark. 'The Atomic Scientists, the Sense of Wonder, and the Bomb', *Environmental History* 12.3 (2007): 578–613.

Fitzmaurice, Andrew. 'The Genealogy of *Terra nullius*', *Australian Historical Studies* 38.129 (2007): 1–15.

Fox, Sarah Alisabeth. *Down Wind: A People's History of the Nuclear West*. Lincoln: University of Nebraska Press, 2014.

Frank, Pat. *Alas, Babylon*. New York: Perennial, 2005. eBook.

Freud, Sigmund. 1971. 'The "Uncanny"' in *Collected Papers*. Trans. Joan Riviere, Vol. 4. London: The Hogarth Press and the Institute of Psychoanalysis.

Furukama, Hideo. *Horses, Horses, in the End the Light Remains Pure: A Tale that begins with Fukushima*. Trans. Doug Slaymaker with Akiko Takenaka. New York: Columbia University Press, 2016.

Gabriel, Ralph. 'Nationalism and the Atom', *Virginia Quarterly Review* 33.4 (1957): 539–48.

Galison, Peter and Alexi Assmus. 'Artificial Clouds, Real Particles', in David Gooding, Trevor Pinch, and Simon Schaffer (eds) *The Uses of Experiment: Studies in the Natural Sciences*. Cambridge: Cambridge University Press, 1989. 225–74.

Gallagher, Carole. 'Nuclear Photography: Making the Invisible Visible', *Bulletin of the Atomic Scientists* 69.6 (2013): 42–6.

Ganetz, Hillevi. 'The Nobel Celebrity-scientist: Genius and Personality', *Celebrity Studies* 7.2 (2016): 234–48.

Ganguly, Debjani. 'Catastrophic Form and Planetary Realism', *New Literary History* 51.2 (2020): 419–453.

George, James. *Ocean Roads*. Wellington: Huia Press, 2006.

Giamo, Benedict. 'The Myth of the Vanquished: The Hiroshima Peace Memorial Museum', *American Quarterly* 55.4 (2003): 703–28.

Gleick, James. *Genius: The Life and Science of Richard Feynman*. New York: Pantheon, 1992. eBook.

Goin, Peter. *Nuclear Landscapes*. Baltimore and London: Johns Hopkins University Press, 1991.

Goldstein, Donald, Katherine Dillon and J. Michael Wenger. *Rain of Ruin: A Photographic History of Hiroshima and Nagasaki*. Dulles, Virginia: Brassey's, 1999.

Goodman, David. Edited and translated. *After Apocalypse: Four Japanese Plays of Hiroshima and Nagasaki*. Ithaca, NY: East Asia Program, 1994.

Gray, Gene. 'Reading the Lisbon Earthquake: Adorno, Lyotard, and the Contemporary Sublime', *Yale Journal of Criticism* 17.1 (2004): 1–18.

Grebowicz, Margret. 'Ecology after Dark: Chernobyl's Wild Horses and the Traffic in Desire', *Minnesota Review* 96 (2021): 56–68.

Greene, Mott. 'Writing Scientific Biography', *Journal of the History of Biology* 40.4 (2007): 727–59.

Griffiths, Alison. '"Shivers Down your Spine": Panoramas and the Origins of the Cinematic Reenactment', *Screen* 41.1 (2003): 1–37.

Groves, Leslie M. *Now It Can be Told: The Story of the Manhattan Project*. Boston: Da Capo Press, 1962. eBook.

Gunn, Joshua and David E. Beard. 'On the Apocalyptic Sublime', *Southern Journal of Communication* 65.4 (2000): 269–86.

Gusterson, Hugh. 'Nuclear Weapons and the Other in the Western Imagination', *Cultural Anthropology* 14.1 (1999): 111–43.

Hachiya, Michihiko. *Hiroshima Diary: The Journal of a Japanese Physician, August 6-September 30, 1945: Fifty Years Later*. Trans. Warner Wells. 1955. Chapel Hill and London: University of North Carolina Press, 1983.

Hales, Peter B. 'The Atomic Sublime', *American Studies* 3.1 (1991): 5–31.

Hankins, Thomas L. 'In Defence of Biography: The Use of Biography in the History of Science', *History of Science* 17.1 (1979): 1–16.

Hawkins, Gay and Stephen Muecke. 'Introduction: Cultural Economies of Waste', in Gay Hawkins and Stephen Muecke (eds) *Culture and Waste: The Creation and Destruction of Value*. Lanham, MD: Rowman & Littlefield, 2003. ix-l.

Hart, Thomas E. 'Mapping Vergil's Quantitative Sublime', *PMLA* 130.3 (2015): 841–3.

Hecht, David K. 'The Atomic Hero: Robert Oppenheimer and the Making of Scientific Icons in the Early Cold War', *Technology and Culture* 49. 4 (2008): 943–66.

———. 'A Nuclear Narrative: Robert Oppenheimer, Autobiography, and Public Authority', *Biography* 33.1 (2010): 167–84.

Hecht, Gabrielle. 'Globalization Meets Frankenstein? Reflections on Terrorism, Nuclearity, and Global Technopolitical Discourse', *History and Technology* 19.1 (2003): 1–8.

———. 'A Cosmogram for Nuclear Things', *Isis* 98.1 (2007): 100–8.

———. 'The Power of Nuclear Things', *Technology and Culture* 51. 1 (2010): 1–30.

Heise, Ursula K. *Sense of Place and Sense of Planet: The Environmental Imagination of the Global*. New York: Oxford University Press, 2008.

——— *Imagining Extinction: The Cultural Meanings of Endangered Species*. Chicago and London: University of Chicago Press, 2016.

Hernandez, Julie Gerk. 'The Tortured Body, the Photograph, and the U.S. War on Terror', *CLC Web: Comparative Literature and Culture* 9.1 (2007). http://dx.doi.org/10.7771/1481-4374.1019.

Hersey, John. *Hiroshima*. New York: Penguin, 2019.

Hesse-Honegger, Cornelia and Peter Wallimann. 'Malformation of True Bug (Heteroptera): A Phenotype Field Study on the Possible Influence of Artificial Low-Level Radioactivity', *Chemistry and Biodiversity* 5 (2008): 499–539.

Higginbotham, Adam. *Midnight in Chernobyl: The Untold Story of the World's Greatest Nuclear Disaster*. New York: Simon & Schuster, 2019.

Hirsch, Marianne. 'Family Pictures: Maus, Mourning, and Post-Memory', *Discourse* 15. 2 (1992–93): 3–29.

———. 'Introduction: Familial Looking', in Marianne Hirsch (ed) *The Familial Gaze*. Hanover: University Press of New England, 1999. xi–xxv.

Hurley, Jessica. 'The Nuclear Uncanny in Oceania', *Commonwealth: Essays and Studies* 41.1 (2018): 95–105. https://journals.openedition.org/ces/396.

Ibuse, Masuji. *Black Rain*. Trans. John Bester. New York: Kodansha, 2012.

If You Love This Planet. Director Terre Nash. National Film Board of Canada. 1982.

Indigenous Environmental Network. 'Indigenous Anti-Nuclear Summit Declaration'. 1996. www.ienearth.org/indigenous-anti-nuclear-summit-decl aration/. 5 April 2022.

Jabiluka. Journeyman Pictures and Frontline Films. 1997.

James Yamazaki. *Children of the Atomic Bomb: An American Physician's Memoir of Nagasaki, Hiroshima, and the Marshall Islands*, 1995.

Japanese Broadcasting Corporation. Ed. *Unforgettable Fire: Pictures Drawn by Atomic Bomb Survivors*. New York: Pantheon, 1977.

Johnson, Rebecca L. *Chernobyl's Wild Kingdom: Life in the Dead Zone*. Minneapolis: Twenty-First Century Books, 2015.

Kaiser, David. 'The Atomic Secret in Red Hands? American Suspicions of Theoretical Physicists During the Early Cold War', *Representations* 90.1 (2005): 28–60.

Källstrand, Gustav. 'Warburg's Dogs: Nobel Laureates and Scientific Celebrity', *Celebrity Studies* 13.1 (2020): 56–72.

Kaplan, E. Ann. *Climate Trauma: Foreseeing the Future in Dystopian Film and Fiction*. New Brunswick, New Jersey: Rutgers University Press, 2016.

Kawakami, Hiromi. 'God Bless You, 2011'. Trans. Ted Goossen and Motoyuki Shibata. *Granta*. 20 March 2012. https://granta.com/god-bless-you-2011/. 3 April 2022.

Keating, Fiona. ' "Remarkable": Scientists amazed by thriving marine life at Bikini Atoll site where 23 atomic bombs were dropped'. *The Independent*. 15 July 2017. www.independent.co.uk/news/science/fish-nuclear-weapons-bombs-sea-stanford-university-us-tests-hiroshima-a7842436.html . 15 March 2022.

Kelly, Cynthia. Ed. *The Manhattan Project: The Birth of the Atom Bomb in the Eyes of Its Creators, Eyewitnesses, and Historians*. New York: Black Dog and Leventhal, 2007.

Keown, Michelle. 'Waves of Destruction: Nuclear Imperialism and Anti-nuclear Protest in the Indigenous Literatures of the Pacific', *Journal of Postcolonial Writing* 54.5 (2018): 585–600.

Kermode, Frank. *The Sense of An Ending: Studies in the Theory of Fiction*. Oxford: Oxford University Press, 2000.

Kirchhof, Astrid Mignon and Jan-Henrik Meyer. 'Global Protest Against Nuclear Power. Transfer and Transnational Exchange in the 1970s and 1980s', *Historical Social Research/Historische Sozialforschung* 39.1 (2014): 165–90.

Kirk, Andrew G. and Kristian Purcell. *Doom Towns: The People and Landscapes of Atomic Testing, A Graphic History*. New York: Oxford University Press, 2016.

Kistiakowsky, George. Quotes from Trinity Test Observers. 16 July 2021. https://armscontrolcenter.org/quotes-from-trinity-test-observers/. 4 May 2022.

Klaver, Elizabeth. *Sites of Autopsy in Contemporary Culture*. Albany: State University of New York Press, 2005.

Kraft, Michael E. and Bruce B. Clary. 'Public Testimony in Nuclear Waste Repository Hearings: A Content Analysis', in Riley E. Dunlap, Michael E. Kraft, and Eugene A. Rosa (eds) *Public Reactions to Nuclear Waste: Citizens' Views of Repository Siting*. Durham: Duke University Press, 1993. 89–114.

Kragh, Helen. 'On Scientific Biography and Biographies of Scientists', in Theodore Arabatzis, Jürgen Renn, Ana Simões (eds) *Relocating the History of Science: Essays in Honor of Kostas Gavroglu*. New York: Springer, 2015. 269–80.

Kyoko and Mark Selden. Ed. *The Atomic Bomb: Voices from Hiroshima and Nagasaki*. London: Routledge, 2015.

Lanouette, William. With Bela Szilard. *Szilard, the Man Behind the Bomb*. New York: Skyhorse, 2013.

Laqueur, Thomas W. *The Work of the Dead: A Cultural History of Mortal Remains*. Princeton: Princeton University Press, 2015.

Laurence, William L. *Dawn Over Zero: The Story of the Atomic Bomb*. 1946. No place: Eschenburg, 2017.

Lee, Edmund W.J. and Shirley S. Ho. 'Are Photographs Worth More Than a Thousand Words? Examining the Effects of Photographic-Textual and Textual-Only Frames on Public Attitude Toward Nuclear Energy and Nanotechnology', *Journalism and Mass Communication Quarterly* 95.4 (2018): 948–70.

Lemmerich, Jost. *Science and Conscience: The Life of James Franck*. Trans. Ann M. Hentschel. Stanford: Stanford University Press, 2011.

Lennon, Jessie. *I'm the One that Know this Country!* Canberra: Aboriginal Studies Press, 2011.

Lifton, Robert Jay and Greg Mitchell. *Hiroshima in America: A Half Century of Denial*. New York: Avon, 1995.

Lindbladh, Johanna. 'Representations of the Chernobyl Catastrophe in Soviet and Post-Soviet Cinema: The Narratives of Apocalypse', *Studies in Eastern European Cinema* 10.3 (2013): 240–56.

Lindee, Susan M. *Suffering Made Real: American Science and the Survivors at Hiroshima*. Chicago: University of Chicago Press, 1994.

———. 'Experimental Wounds: Science and Violence in Mid-Century America', *Journal of Law, Medicine, and Ethics* 39.1 (2011): 8–20.

Lindholm, Charles. 'The Anthropology of Charisma', in José Pedro Zúquete (ed) *Routledge International Handbook of Charisma*. Oxford: Routledge, 2021. 39–49.

Lippit, Akira Mizuta. *Atomic Light (Shadow Optics)*. Minneapolis: University of Minnesota Press.

Lury, Celia. '"Bringing the World into the World": The Material Semiotics of Contemporary Culture', *Distinktion: Scandinavian Journal of Social Theory* 13.3 (2012): 247–60.

Lynch, Lisa. '"We Don't Wanna Be Radiated": Documentary Film and the Evolving Rhetoric of Nuclear Energy Activism', *American Literature* 84.2 (2012): 327–51.

Makki, Fouad. 'Development by Dispossession: *Terra Nullius* and the Social-Ecology of New Enclosures in Ethiopia', *Rural Sociology* 79.1 (2014): 79–103.

Makkreel, Rudolf. 'Imagination and Temporality in Kant's Theory of the Sublime', *Journal of Aesthetics and Art Criticism* 42.3 (1984): 303–15.

Manhattan Engineer District. *Photographs of the Atomic Bombings of Hiroshima and Nagasaki*. Parts 1-3. Manhattan Engineer District, 1945.

Marcoń, Barbara. 'Hiroshima and Nagasaki in the Eye of the Camera', *Third Text* 25.6 (2011): 787–97.

Marder, Michael and Anaïs Tondeur. *The Chernobyl Herbarium: Fragments of an Exploded Consciousness*. London: Open Universities Press, 2016.

Marples, Megan and Rachel Ramirez. 'The Doomsday Clock Reveals How Close We Are to...Doom'. *CNN*. 20 January 2022. https://edition.cnn.com/2022/01/20/world/doomsday-clock-2022-climate-scn/index.html. 30 March 2022.

Marshall, P. David, Christopher Moore and Kim Barbour. 'Persona as Method: Exploring Celebrity and the Public Self through Persona Studies', *Celebrity Studies* 6.3 (2015): 288–305.

Masco, Joseph. 'Nuclear Technoaesthetics: Sensory Politics from Trinity to the Virtual Bomb in Los Alamos', *American Ethnologist* 31. 3 (2004): 1–25.

———. *The Nuclear Borderlands: The Manhattan Project in Post-Cold War New Mexico*. Princeton, NJ: Princeton University Press, 2006.

———. 'Engineering the Future as Nuclear Ruin', in Ann Laura Stoler (ed) *Imperial Debris: On Ruins and Ruination*. Durham and London: Duke University Press, 2013. 252–86.

Matashichi, Ōishi. *The Day the Sun Rose in the West: Bikini, the Lucky Dragon, and I*. Trans. Richard H. Minear. Honolulu: University of Hawai'i Press, 2011.

McMillan, Priscilla. *The Ruin of J. Robert Oppenheimer and the Birth of the Modern Arms Race*. New York: Viking, 2005.

Medvedev, Grigoriy. *Chernobyl Notebook*. 1987. Kindle Edition.

Meeuf, Russell. 'Nuclear Epistemology: Apocalypticism, Knowledge, and the "Nuclear Uncanny" in *Kiss Me Deadly*', *LIT: Literature Interpretation Theory* 23.3 (2012): 283–304.

Miller, Alyson and Cassandra Atherton. ' "Monster in the Sky": Hibakusha Poetry and the Nuclear Sublime', *Text* 41 (2017): 1–12. www.textjournal.com.au/speciss/issue41/Miller&Atherton.pdf

Minear, Richard H. 'Review: The Atomic-bomb Paintings', *Bulletin of Concerned Asian Scholars* 19.4 (1987): 58–63.

———. Ed. and Trans. *Hiroshima: Three Witnesses*. Princeton, NJ: Princeton University Press, 1990.

Mitchell, W.J.T. 'Image, Space, Revolution: The Arts of Occupation', *Critical Inquiry* 39.1 (2012): 8–32.

Mizue, Akiko. *Hiroshima: Survivors' Testimonies*. NP: Ittosha Inc., 2020. eBook.

Monk, Ray. *Inside the Center: The Life of J. Robert Oppenheimer*. New York: Random House, 2012.

Mooney, Annabelle. *Human Rights and the Body: Hidden in Plain Sight*. Farnham, Surrey: Ashgate, 2014.

Morrison, Philip. 'If the Bomb Gets Out of Hand', in Dexter Masters and Katharine Way (eds) *One World or None*. New York: McGraw-Hill, 1946. 1–6.

Morrison, Susan Signe. *The Literature of Waste: Material Ecopoetics and Ethical Matter*. New York: Palgrave Macmillan, 2015.

Morton, Tim. *Hyperobjects: Philosophy and Ecology after the End of the World*. Minneapolis: University of Minnesota Press, 2013.

Moss, Jim. 'The Concrete Sublime', *Architectural Theory Review* 18.2 (2013): 251–7.

Nagai, Takashi. *The Bells of Nagasaki*. Trans. William Johnston. Tokyo: Kodansha International, 1984.

Nakazawa, Keiji. *Barefoot Gen. Vol. I*. Trans. Project Gen. San Francisco: Last Gasp, 2004.

———. *Barefoot Gen. Vol. II: The Day After*. Trans. Project Gen. San Francisco: Last Gasp, 2004.

———. *Barefoot Gen. Vol. III: Life After the Bomb*. Trans. Project Gen. Philadelphia: New Society, 1994.

———. *Barefoot Gen. Vol. IV: Out of the Ashes*. Trans. Project Gen. San Francisco: Last Gasp, 2005.

———. *Barefoot Gen. Vol. V: The Never-Ending War*. Trans. Project Gen. San Francisco: Last Gasp, 2007.

———. *Barefoot Gen. Vol. VI: Writing the Truth*. Trans. Project Gen. San Francisco: Last Gasp, 2007.

———. *Barefoot Gen. Vol. VII: Bones into Dust*. Trans. Project Gen. San Francisco: Last Gasp, 2008.

———. *Barefoot Gen. Vol. IX: Breaking Down Boundaries*. Trans. Project Gen. San Francisco: Last Gasp, 2018.

———. *Barefoot Gen. Vol X: Never Give Up*. Trans. Project Gen. San Francisco: Last Gasp, 2009.

Nancy, Jean-Luc. *After Fukushima: The Equivalence of Catastrophes*. Trans. Charlotte Mandel. New York: Fordham University Press, 2015.

National Committee on Atomic Information and Philip Ragan Productions. *One World or None*. Library of Congress, 1946.

Nayar, Pramod K. *Ecoprecarity: Vulnerable Lives in Literature and Culture*. New York: Routledge, 2019.

———. *Bhopal's Ecological Gothic: Disaster, Precarity, and the Biopolitical Uncanny*. Maryland: Lexington, 2017.

Niskanen, Kristi, Mineke Bosch and Kaat Wils. 'Scientific Personas in Theory and Practice – Ways of Creating Scientific, Scholarly, and Artistic Identities', *Persona Studies* 4.1 (2018): 1–5.

Nuclear Wasteland. Director Timothy Large. Thomson Reuters Foundation.

Oe, Kenzaburo. *Hiroshima Notes*. Trans. David L. Swain and Toshi Yonezawa. New York: Grove, 1981.

Okamura, Yukinori. 'The Hiroshima Panels Visualize Violence: Imagination Over Life', *Journal for Peace and Nuclear Disarmament* 2.2 (2019): 518–34.

Oliver, Kelly. 'Witnessing, Recognition, and Response Ethic', *Philosophy & Rhetoric* 48.4 (2015): 473–93.

Oldfield, Samantha-Jayne. 'Narrative Methods in Sport History Research: Biography, Collective Biography, and Prosopography', *The International Journal of the History of Sport* 32.15 (2015): 1855–82.

Oppenheimer, J.R. 'The New Weapon: The Turn of the Screw', in Dexter Masters and Katharine Way (eds) *One World or None*. New York: McGraw-Hill, 1946. 22–5.

———. 'Oppenheimer on Einstein', *Bulletin of the Atomic Scientists* 35.3 (1979): 36–9.

Osada, Arata. Compiled. *Children of the A-Bomb: Testament of the Boys and Girls of Hiroshima*. Trans. Jean Dan and Ruth Sieben-Morgan. Auckland: Pickle Partners, 2015. eBook.

Ottaviani, Jim and Leland Myrick. *Feynman*. New York: First Second, 2011.

———., Janine Johnston, Steve Lieber, Vince Locke, Bernie Mireault and Jeff Parker. *Fallout: J. Robert Oppenheimer, Leo Szilard and the Political Science the Atom Bomb*. GT Labs, 2013.

Outram, Dorinda. 'Life-paths: Autobiography, Science and the French Revolution', in Michael Shortland and Richard Yeo (eds) *Telling Lives in Science: Essays on Scientific Biography*. Cambridge: Cambridge University Press, 1996. 85–102.

Pais, Abraham, J. *Robert Oppenheimer: A Life*. Oxford: Oxford University Press, 2006.

Pajo, Judi. 'Two Paradigmatic Waves of Public Discourse on Nuclear Waste in the United States, 1945-2009: Understanding a Magnitudinal and Longitudinal Phenomenon in Anthropological Terms', *PLoS One* 11.6 (2016): e0157652. doi:10.1371/journal.pone.0157652.

Paul, Herman. 'What is a Scholarly Persona: Ten Theses on Virtues, Skills and Desires', *History and Theory* 53 (2014): 348–71.

———. 'Sources of the Self. Scholarly Personae as Repertoires of Scholarly Selfhood', *BMGN – Low Countries Historical Review* 131.4 (2016): 135–54.

Pauli, Hertha. 'Nobel's Prizes and the Atom Bomb'. *Commentary*. December 1945. www.commentary.org/articles/hertha-pauli/nobels-prizes-and-the-atom-bomb/ 19 December 2021.

Peeples, Jennifer. 'Toxic Sublime: Imaging Contaminated Landscapes', *Environmental Communication* 5.4 (2011): 373–92.

Penny, Sara. 'Personal Account of Downwinders and Nuclear Weapons Testing'. Undated. https://collections.lib.utah.edu/details?id=1251771. 3 April 2022.

Plokhy, Serhii. *Chernobyl: The History of a Nuclear Catastrophe*. New York: Basic, 2018.

Presto, Jennifer. 'The Aesthetics of Disaster: Blok, Messina, and the Decadent Sublime', *Slavic Review* 70.3 (2011): 569–90.

Pringle, Thomas. 'Photographed by the Earth: War and Media in Light of Nuclear Events', *European Journal of Media Studies* 3.2 (2014): 131–54.

Radomska, Marietta and Cecilia Åsberg. 'Doing Away with Life: On Biophilosophy, the Non/Living, Toxic Embodiment, and Reimagining Ethics', in Erich Berger, Kasperi Mäki-Reinikka, Kira O'Reilly, Helena Sederholm (eds) *Art as We Don't Know It*. Aalto University, 2020. 52–61

Redniss, Lauren. *Radioactive: Marie and Pierre Curie: A Tale of Love and Fallout*. New York: Day Street, 2011.

Rhodes, Richard. *The Making of the Atomic Bomb*. London: Simon and Schuster, 2012.

Richie, Donald. '*Mono no Aware*: Hiroshima in Film', in Mick Broderick (ed) *Hibakusha Cinema: Hiroshima, Nagasaki and the Nuclear Image in Japanese Cinema*. Routledge, 2009. eBook. Unpaginated.

Richter, Darmon. *Chernobyl: A Stalkers' Guide*. London: FUEL, 2020.

Rifkind, Candida. 'The Seeing I of Scientific Graphic Biography', *Biography* 38.1 (2015): 1–22.

Roberts, Annelise. '"Knowing This Country": Confronting the Nuclear Uncanny in Aboriginal Life Writing', *Journal of the Association for the Study of Australian Literature* 21.2 (2021): 1–13.

Rose, Deborah Bird, Thom van Dooren, and Matthew Chrulew. Eds. *Extinction Studies: Stories of Time, Death, and Generations*. New York: Columbia University Press, 2017.

Roy, Arundhati. 'The End of Imagination', in *The Algebra of Infinite Justice*. New Delhi: Penguin, 2002. 1–42.

Russell-Einstein Manifesto. 1955. Archived at www.atomicheritage.org/key-documents/russell-einstein-manifesto. 13 March 2022.

Saint-Amour, Paul. 'Bombing and the Symptom: Traumatic Earliness and the Nuclear Uncanny', *Diacritics* 30.4 (2000): 59–82.

Sawyer, Dylan. *Lyotard, Literature and the Trauma of the Differend*. Basingstoke: Palgrave-Macmillan, 2014.

Schroeder, Robyn. 'The Rise of the Public Humanities', in Susan Smulyan (ed) *Doing Public Humanities*. New York: Routledge, 2021. 5–27.

Schweber, Silvan S. *In the Shadow of the Bomb: Oppenheimer, Bethe, and the Moral Responsibility of the Scientist*. Princeton, NJ: Princeton University Press, 2000.

———. *Einstein and Oppenheimer: The Meaning of Genius*. Cambridge, MA: Harvard University Press, 2010.

———. *Nuclear Forces: The Making of the Physicist Hans Bethe*. Cambridge, MA: Harvard University Press, 2012.

Schuppli, Susan. *Material Witness: Media, Forensics, Evidence*. Cambridge, MA: MIT, 2020.

Segrè, Gino and Bettina Hoerlin. *The Pope of Physics: Enrico Fermi and the Birth of the Atomic Age*. New York: Henry Holt, 2016.

Sen, Udita. 'Developing *Terra Nullius*: Colonialism, Nationalism, and Indigeneity in the Andaman Islands', *Comparative Studies in Society and History* 59.4 (2017): 944–73.

Shapin, Steven. *A Social History of Truth. Civility and Science in Seventeenth-Century England*. Chicago and London: University of Chicago Press, 1994.

———. *The Scientific Life: A Moral History of a Late Modern Vocation*. Chicago: University of Chicago Press, 2008.

——— and Arnold Thackray. 'Prosopography as a Research Tool in History of Science: The British Scientific Community, 1700-1900', *History of Science* 12 (1974): 1–28.

Shapiro, Michael J. *The Political Sublime*. Durham and London: Duke University Press, 2018.

Shibata, Yuko. *Producing Hiroshima and Nagasaki: Literature, Film, and Transnational Politics*. Honolulu: University of Hawai'i Press, 2018.

Shields, Lora M and Philip V. Wells. 'Effects of Nuclear Testing on Desert Vegetation', *Science* New Series 135.3497 (1962): 38–40.

Shindell, Matthew. *The Life and Science of Harold C. Urey*. Chicago: University of Chicago Press, 2020.

Shute, Nevil. *On the Beach*. Pearson Education, 2008. eBook.

Smith, Alice Kemball. *A Peril and a Hope: The Scientists' Movement in America 1945-1947*. Cambridge: MIT Press, 1971.

Smulyan, Susan. 'What Can Public Art Teach Public Humanities', in Susan Smulyan (ed) *Doing Public Humanities*. New York: Routledge, 2021. 28–38.

Smyth, Henry deWolf. *Atomic Energy for Military Purposes: The Official Report on the Development of the Atom Bomb under the Auspices of the United States Government, 1940-1945*. Princeton: Princeton University Press, 1948.

Söderqvist, Thomas. 'Existential Projects and Existential Choice in Science: Science Biography as an Edifying Genre', in Michael Shortland and Richard

Yeo (eds) *Telling Lives in Science: Essays on Scientific Biography*. Cambridge: Cambridge University Press, 1996. 45–84.

Solnit, Rebecca. *Savage Dreams: A Journey Into the Hidden Wars of the American West*. Berkeley, London: University of California Press, 2014. eBook.

Southard, Susan. *Nagasaki: Life After Nuclear War*. New York: Viking, 2015.

Sperling, Alison. 'Radiating Exposures', in Christoph F.E. Holzhey and Arnd Wedemeyer (eds) *Weathering: Ecologies of Exposure, Cultural Inquiry* 17 (2020): 41–62.

Spitz, Chantal. *Island of Shattered Dreams*. Hula, 2013.

Stephens, Lloyd and Rosemary Barrett, 'A Brief History of a 20th Century Danger Sign'. *Lawrence Berkeley National Laboratory*. LBNL Report #: LBL-7260. https://escholarship.org/uc/item/7cz9p0m6. 16 March 2022.

Sturges, R.P. 'From Collected Biography to Prosopography', *Library Review* 25.5–6 (1976): 210–3.

Tatsuta, Kazuto. *ICHI-F: A Worker's Graphic Memoir of the Fukushima Nuclear Power Plant*. New York: Kodansha, 2017.

Tawada, Yoko. *The Last Children of Tokyo*. Trans. Margaret Mistutani. London and New York: Portobello-New Directions, 2017.

Taylor Bryan C. and Judith Hendry. 'Insisting on Persisting: The Nuclear Rhetoric of "Stockpile Stewardship"', *Rhetoric & Public Affairs* 11.2 (2008): 303–34.

Taylor, N.A.J. and Robert Jacobs. Eds. *Reimagining Hiroshima and Nagasaki: Nuclear Humanities in the Post-Cold War*. Oxford: Routledge, 2018.

The Beginning of the End of the Nuclear Weapons. Director Álvaro Orús. 2019.

The Rand Corporation. *Worldwide Effects of Atomic Weapons: Project Sunshine*. Santa Monica: The Rand Corporation, 1956.

Thompson, Helen. 'Chernobyl's Bugs: The Art And Science Of Life After Nuclear Fallout'. *Smithsonian Magazine*. 26 April 2014. www.smithsonianmag.com/arts-culture/chernobyls-bugs-art-and-science-life-after-nuclear-fallout-180951231/. 15 March 2022.

Thomson, Rosemarie Garland. *Extraordinary Bodies: Figuring Physical Disability in American Culture and Literature*. New York: Columbia University Press, 1997.

Thorpe, Charles. 'Disciplining Experts: Scientific Authority and Liberal Democracy in the Oppenheimer Case', *Social Studies of Science* 32.4 (2002): 525–62.

———. *Oppenheimer: The Tragic Intellect*. Chicago and London: University of Chicago Press, 2006.

——— and Steven Shapin. 'Who Was J. Robert Oppenheimer? Charisma and Complex Organization', *Social Studies of Science* 30.4 (2000): 545–90.

Treat, John Whittier. *Writing Ground Zero: Japanese Literature and the Atomic Bomb*. Chicago and London: University of Chicago Press, 1996.

Trigg, Dylan. *The Memory of Place: A Phenomenology of the Uncanny*. Athens: Ohio University Press, 2012.

Turvey, Samuel T. and Anthony S. Cheke. 'Dead as a Dodo: The Fortuitous Rise to Fame of an Extinction Icon', *Historical Biology* 20.2 (2008): 149–63.

Tuwhare, Hone. 'No Ordinary Sun', in Hone Tuwhare (ed), *No Ordinary Sun: Poems*. Auckland: Blackwood and J. Paul, 1965. 23.

Tynan, Elizabeth. *Atomic Thunder: The Maralinga Story*. Sydney: University of New South Wales Press, 2016.

United States Atomic Energy Commission. *In the Matter of J. Robert Oppenheimer: Texts of Principal Documents and Letters*. Cambridge: MIT Press, 1971.

Uranium: Is It a Country? Director. Kerstin Schnatz. Berlin, Germany: Initiative Strahlendes Klima, 2008.

Urey, Harold C. 'How Does it All Add Up?' in Dexter Masters and Katharine Way (eds) *One World Or None*. New York: McGraw-Hill, 1946. 53–8.

Verboven, Koenraad, Myriam Carlier and Jan Dumolyn. 'A Short Manual to the Art of Prosopography'. https://prosopography.history.ox.ac.uk/images/01%20Verboven%20pdf.pdf

Virilio, Paul. *War and Cinema: The Logistics of Perception*. Trans. Patrick Camiller. London: Verso, 1984.

Wake, Naoko. *American Survivors: Trans-Pacific Memories of Hiroshima and Nagasaki*. Cambridge: Cambridge University Press, 2021.

Walker, Cheryl. 'Persona Criticism and the Death of the Author', in William H. Epstein (ed) *Contesting the Subject: Essays in the Postmodern Theory and Practice of Biography and Biographical Criticism*. West Lafayette: Purdue University Press, 1991. 109–21.

'War's Ending: Atomic Bomb and Soviet Entry Bring Jap Surrender Offer', *Life* 20 August 1945. 25–31.

Weart, Spencer R. *The Rise of Nuclear Fear*. Cambridge, MA: Harvard University Press, 2012.

Weisman, Alen. *The World Without Us*. New York: Thomas Dunne, 2007.

Weller, George. *First into Nagasaki: The Censored Eyewitness Dispatches on Post-Atomic Japan and Its Prisoners of War*. Edited by Anthony Weller. New York: Crown, 2016.

Wellerstein, Alex. *Restricted Data: The History of Nuclear Secrecy in the United States*. Chicago: University of Chicago Press, 2021.

Wells, H.G. *The World Set Free*. eBook.

White, Hayden. 'Figural Realism in Witness Literature', *Parallax* 10.1 (2004): 113–24.

White, Paul. 'Darwin, Concepción, and the Geological Sublime', *Science in Context* 25.1 (2012): 49–71.

Wilson, Rob. 'Towards the Nuclear Sublime: Representations of Technological Vastness in Postmodern American Poetry', *Prospects* 14 (1989): 407–39.

Woodall, Joanna. 'Laying the Table: The Procedures of Still Life', *Art History* 35.5 (2012): 976–1003.

Wright, Patrick. 'A Timeless Sublime?' *Angelaki* 15.2 (2010): 85–100.

Wylie, Philip. *Tomorrow!* New York and Toronto: Rinehart and Company, 1954. eBook.

Wyndham, John. *Re-birth*. New York: Ballantine Books, 2013. eBook.

Yaeger, Patricia. 'Trash as Archive, Trash as Enlightenment', in Gay Hawkins and Stephen Muecke (eds) *Culture and Waste: The Creation and Destruction of Value*. Lanham, MD: Rowman & Littlefield, 2003. 103–16.

Yoneyama, Lisa. *Hiroshima Traces: Time, Space and the Dialectics of Memory*. Berkeley and London: University of California Press, 1999.

Zabytko, Irene. *The Sky Unwashed*. Chapel Hill: Algonquin Books, 2012. eBook.

Zwigenberg, Ray. *Hiroshima: The Origins of Global Memory Culture*. Cambridge: Cambridge University Press, 2014.

Index